how to
know the
mites
and ticks

The **Pictured Key Nature Series** has been published since 1944 by the Wm. C. Brown Company. The series was initiated in 1937 by the late Dr. H. E. Jaques, Professor Emeritus of Biology at Iowa Wesleyan University. Dr. Jaques' dedication to the interest of nature lovers in every walk of life has resulted in the prominent place this series fills for all who wonder **"How to Know."**

John F. Bamrick and Edward T. Cawley
Consulting Editors

The Pictured Key Nature Series

How to Know the
AQUATIC INSECTS, Lehmkuhl
AQUATIC PLANTS, Prescott
BEETLES, Arnett-Downie-Jaques, Second Edition
BUTTERFLIES, Ehrlich
ECONOMIC PLANTS, Jaques, Second Edition
FALL FLOWERS, Cuthbert
FERNS AND FERN ALLIES, Mickel
FRESHWATER ALGAE, Prescott, Third Edition
FRESHWATER FISHES, Eddy-Underhill, Third Edition
GILLED MUSHROOMS, Smith-Smith-Weber
GRASSES, Pohl, Third Edition
IMMATURE INSECTS, Chu
INSECTS, Bland-Jaques, Third Edition
LAND BIRDS, Jaques
LICHENS, Hale, Second Edition
LIVING THINGS, Jaques, Second Edition
MAMMALS, Booth, Third Edition
MARINE ISOPOD CRUSTACEANS, Schultz
MITES AND TICKS, McDaniel
MOSSES AND LIVERWORTS, Conard-Redfearn, Third Edition
NON-GILLED FLESHY FUNGI, Smith-Smith
PLANT FAMILIES, Jaques
POLLEN AND SPORES, Kapp
PROTOZOA, Jahn, Bovee, Jahn, Third Edition
SEAWEEDS, Abbott-Dawson, Second Edition
SEED PLANTS, Cronquist
SPIDERS, Kaston, Third Edition
SPRING FLOWERS, Cuthbert, Second Edition
TREMATODES, Schell
TREES, Miller-Jaques, Third Edition
TRUE BUGS, Slater-Baranowski
WATER BIRDS, Jaques-Ollivier
WEEDS, Wilkinson-Jaques, Third Edition
WESTERN TREES, Baerg, Second Edition

how to
know the
mites
and ticks

Burruss McDaniel
South Dakota State University

The Pictured Key Nature Series
Wm. C. Brown Company Publishers
Dubuque, Iowa

595.42
MJ34

Contents

Preface

The chief aim of this book is to present simply and straight forwardly the basic facts about mites and ticks collected in the United States. A desire of the author is to arouse in the reader a lasting curiosity to know more about these tiny creatures. This comprehensive illustrated work was undertaken to supplement the many excellent but highly technical publications currently available treating the mites and ticks of the United States. Most students searching for information on these tiny creatures cannot assemble a library large enough to include the many works published nor are they in an area where there is a library that would contain the books and journals needed.

This book is directed toward the general student as well as the specialist. Collecting and mounting techniques are included. The general student will find the many illustrations helpful in placing a specimen within a family and many times, within its genus, simply by seeing the illustrations and making a comparison. For the specialist, the host records and distribution data were obtained from original published papers. The vast majority of the illustrations were drawn from actual specimens and from the illustrations of the original description of that species.

This is by no means a complete list of the mites and ticks of the United States. How-ever, in certain groups it will be complete due to the small number of species described to date. This book is not limited to workers in the United States since many specimens found in this country are also present in other countries throughout the world. To illustrate, the author has found that he is able to use published works from South Africa as an aid in the identification of material collected in the United States. This is not only true for those forms that are parasitic on animals and plants but applies equally well to the soil inhabiting mite fauna. It is understood that the general student will not have a previous knowledge of this subject. Therefore, the general morphological characteristics of each group have been described along with a glossary as an aid in handling the technical language. It is invisioned that a worker would also avail themselves of G. W. Krantz's *Manual of Acarology* as an aid in keying to the family level and then turning to this book to determine the genus, and in many cases, even the species name.

This work is divided into five major divisions with each division constituting a unit in itself. The material is arranged according to the scheme of classification which the author found most generally acceptable to acarological workers. The main section is the treatment of families, genera and selected species. There is an analytical key for each of the families within the five major divisions. However, in many families it was not possible to construct an analytical key and in such cases the genera are listed as 1, 2, 3 etc. with each couplet containing the outstanding morphological structure of that genus. Because of page limitations, many genera are represented by a single species and do not show the variation that normally exists within the genus. However, species were selected which, in the opinion of the author, most represented the generic concept.

It is the desire of the author to create a feeling of concern and goodwill toward mites and ticks in general. The more one knows about a subject the less one has to fear as familiarity breeds a feeling of security. This book will allow the reader to become aware of the innumerable harmless species as well as those species that are detrimental to the economic welfare of man.

In the preparation of a book of this sort, the author is dependent upon the work of a great number of people, so many that it is impossible to thank them all in print. Acknowledgment must be made here, however, of the valuable assistance given by Drs. Bamrick and Cawley of Loras College, Dubuque, Iowa and Editor Louise Barrett and other members of the staff of Wm. C. Brown Company whose expert advise and guidance made the production of this book possible.

The pen and ink drawings were prepared by the author with the help of Mary Miller and later by Cynthia Morihara. Many of the illustrations have been redrawn and adapted from various sources in the literature. My final and deepest appreciation goes to my wife, Shirley, for her patient help in typing the entire manuscript, proofreading, and compiling the bibliography, index and glossary.

Burruss McDaniel

Introducing the Mites and Ticks

The microworld of ticks and mites can be one of the most fascinating adventures of a lifetime. Within the boundaries of this world there exists zoological curiosities that would top the wildest fiction story ever written about creatures from outer space. The parasitic batmites of the family Chirodiscidae are strange forms possessing structural modifications of their legs enabling them to clasp the hair of their host. Members of this family also have the habit of retaining the young within the body of the female, where it is nourished until ready to emerge as a nymphal form.

Throughout the entire series of these small creatures there are phenomena of special interest from the point of view of evolution. The extraordinary development of both structure and habit occurring among ticks and mites offer an endless variety which has barely been explored. There are beautiful examples of what appears to be convergent evolution which has resulted in the development of similar structures and habits in quite unrelated forms. It has been demonstrated that beetles play an important role in the dispersion of mites found in cow dung. Many scarabaeid and larger hydrophilid beetles have been collected with mites riding on them. The mites have been observed on the elytra and ventral surface of the beetle while it was in motion, and remained relatively motionless during this period. When the beetle stopped, the mites moved off and on the beetle continuously. When the beetle moved again, some of the mites remain on the dung material and merely climb onto the next beetle that stopped in their vicinity. This association is an important factor in the dispersion and ecology of several mite species of the family Macrochelidae.

Ticks and mites constitute the major ectoparasitic fauna of the vertebrates. They have been reported as penetrating the lungs, nose, ears, skin, intestine and bladder of vertebrates. They act as transmitting agents of numerous diseases, including protozoa, rickettsia and viruses. They cause considerable discomfort to man and other animals due to their biting and blood-sucking habits. Two very important rickettsial diseases that utilize the ticks and mites as vectors are scrub typhus and Rocky Mountain spotted fever. The former is carried by trombiculid mites called chiggers; the latter is carried by members of the family Ixodidae commonly called ticks. Scrub typhus is a disease of the Asiatic-Pacific region, whereas Rocky Mountain spotted fever has been reported from the United States, Canada, Mexico, Panama and parts of South America.

1

Collecting and Mounting Mites and Ticks

To collect all types of mites, one could search everywhere. Usually students limit themselves to one group. The keeping of records with each collection is of utmost importance. One should have a notebook in which are recorded date, locality, collector's name and any other pertinent information such as type of soil, humidity and temperature. Each collection must be clearly marked to correlate with the notebook entry.

Many mites are associated with plants. For example, the Tetranychoidea (spider mites) occur in colonies on the undersides of leaves, while the smaller Eriophyoidea (gall mites) and Tarsonemoidea are more secretive, hiding out in buds, galls, in crevices on the stem and along the veins on the underside of leaves. Other mites, such as the Phytoseiidae, associate themselves with the leaves and bark of deciduous trees.

There are many complicated ways to collect mites associated with plants but the simple, direct methods of sweeping, beating, aspirating, or wiping the mites off with a moist brush are quite effective. Usually it is best to bring the leaves, stems, etc. to your dissecting microscope where you can pick them off and drop them immediately into a vial of 85% alcohol. It is best to place the plant or parts of plants into a plastic bag along with a notation of collecting data and securely close the bag so as to prevent any mites from crawling out.

A suction method has been developed for extracting Arthropoda from grasslands and any low-growing plants. This method and the apparatus required is thoroughly discussed in Johnson et al. (1957).

The gall-forming eriophyids may be collected by carefully opening the galls and placing them in drying jars or cylinders that have been wiped with glycerol inside the mouth. This prevents the mites from crawling out of the jar or cylinder. Not more than 1 inch of plant material should be placed in the bottom of the cylinder. The drying process consists of placing the cylinder in a dry, airy place protected from direct light. In about 12 hours the mites should be crawling along the wall of the cylinder. Chloropicric acid (a saturated solution of picric acid in 2% hydrochloric acid) should be slightly warmed and poured over the dried plant material and shaken vigorously. Then the cylinder is allowed to stand while the plant material settles to the bottom and the liquid with the mites comes to the top. The latter should be poured off into a suitable container and stored. Other methods of collecting Eriophyid mites are discussed in detail by Keifer (1952).

Parasitic ticks and mites are bountiful since just about any animal larger than they themselves is parasitized by them. Terrestrial, and aquatic animals may serve as hosts, thus it is important to identify the host properly before identifying its parasites. When looking for external parasites from dead birds and other small animals, it is always wise to place the animal in a tightly closed plastic bag to transport it to the laboratory. Those ticks and mites living as external parasites of animals can be collected by simply brushing and combing the body over a large shallow pan. It is usually recommended that the external parasites be killed before removing them from the host. This is best accomplished by placing the dead animal in a plastic bag, inserting a chloroform-soaked wad of cotton and tying the bag shut securely.

External parasites may also be washed from an animal by placing approximately 3 drops of ordinary dishwashing liquid detergent in a wide-mouthed gallon jug half full of water. The animal is dunked up and down several times and then returned to its original plastic bag. The water-detergent mixture is then suctioned off into a petri dish and examined for the presence of ticks and mites. All parasites should be picked out with tweezers and stored in small alcohol-filled vials. The water is then discarded and another portion of the water-detergent mixture is ready to be examined under the dissecting microscope. This method is slow and time consuming but it is quite effective.

When collecting parasites from birds it is always wise to examine the quills of the feathers, the scales of the legs and nasal passages. In insects, the wings, and the elytra of beetles needs to be unfolded. Before examining any host for internal parasites, the ear, eye, anal and genital openings should be thoroughly checked. To date, very few mites have been found in the digestive tract, but many are found in the respiratory system.

Ticks can be readily collected individually from the host by brushing the parasites with glycerol or paraffin before detaching them from the host. When the ticks are loosened in this fashion they can be removed without damaging the hypostome, an important taxonomic structure. Ticks found on vegetation are readily collected by flagging, which consists of trailing a woollen blanket over the infested area.

A word of caution should be stated concerning collection sites of ticks and mites. Most specimens prefer soil not exposed to direct sunlight or wind and one with enough moisture to maintain a satisfactory water balance. Many ticks or other ectoparasites of vertebrates are most likely found in nesting areas, runways or water routes of their hosts.

When examining material in the field, place a small section of soil or debris on a black or white enamel tray. Pick through the section slowly and carefully. You will find that many mites will fall onto the tray's shiny, smooth surface. These should be guided away from the main section of debris and after free of any litter, picked up with a camel's hair brush and placed in a vial containing 85% alcohol. Every effort should be made not to add debris to the vial, especially when soil is virtually impossible to remove.

The tray referred to above can also be used in another way. Some mites, especially chiggers, are curious and investigate any new object in their environment. Therefore, simply leaving the clean tray on the ground for many minutes will frequently bring specimens that can be picked up and placed in a vial. Other collectors recommend "dropping" the tray on the ground as this creates a vibration which activates quiescent Trombiculid larvae.

Free-living parasites, such as those found in nests, can be collected by simple heat-dessication techniques. This technique will also work for collecting other small species in soil, litter and other organic detritus. Berlese (1905) developed the basic funnel-desiccation method and this was revised by

Tullgren (1918). In this heat-desiccation method, the 40W bulb is not only the source of heat, but serves as an extra stimulus to downward movement as most of the animals are negatively phototrophic. The parasites fall through the screen, down the funnel and into the jar or container of 85% alcohol. It will be necessary to sort the various specimens under a dissecting microscope.

Free-living water mites are usually found in the weedy areas around small lakes, streams, and ponds. They may be collected with a small dip net having a narrow mesh bag, a small silk stocking net, or a tea strainer on a long pole. Since the larvae of water mites are frequently parasites, the wings and bodies of beetles, hemipterans, dragonflies, etc. should be examined for the presence of sac-like opaque white or bright orange nymphs in larval skins. A white porcelain tray is best for examining water mites after which they can be picked up with an eye dropper.

The identification of ticks requires that all areas of the specimen must be examined, dorsal, ventral, and lateral regions. In some groups such as the genus *Argas*, a soft tick genus, specimens must be placed in a lateral position in order to detect the presence of the sutural line that characterizes the members of this genus. In the hard ticks the spiracle must be studied, the length of the coxal spurs and number of recurved teeth on the hypostome. To observe all these structures with ease it has been found that the standard Syracuse watch glass turned upside down will enable a worker to orientate the specimen in all positions needed for identification. This is accomplished by using the curved outer depression of the watch glass. In this depression the tick can be maneuvered into various positions that allows the worker using a good dissecting microscope and adequate lighting to see all structures used in tick identification. This same Syracuse watch glass can also be used in dissection work if the hypostome requires removal to view its essential characters clearly.

A distinct advantage in storing ticks in 70-80 percent alcohol is that the specimen will dry quickly thus making surface characters visible. Fluids containing glycerine have the disadvantage of leaving the surface of the tick with a film that masks many of the essential characters. For color preservation carbon tetrachloride can be used, however, this material makes the specimens very hard and brittle.

The vital importance of labeling specimens is here emphasized. A specimen is of no scientific value unless it contains data stored with the material. The data should include the locality, giving a reference point such as a town, road, river, etc., the state and country, date of collection, altitude, host (scientific name and common name or accession number), collector's name and site of attachment on body of host animal or plant.

How to Use the Keys

Identification rests upon a selection of comparatively obvious characters that may be seen in the specimens. The mere shape or outline can facilitate identification. The student who attempts to identify an unknown North American mite or tick with the aid of keys should have a knowledge in the use of the microscope. An adequate microscope preparation is the only way in which positive identification of mites can be made.

The intention of the included keys is to make possible the identification of a single complete adult male and female. Other information that may be utilized is the style of living, i.e., free-living or parasitic.

The use of a key in identification is the first major obstacle one encounters in attempting to place a name to a specific animal. First of all, it is important to read through the key entirely at least once until you have a good idea of the way it is organized. The structure of a particular key will dictate how the key best serves the worker. In some keys, several characters are given as alternates to the main character that is supposed to direct the worker to the next cuplet. Alternates in which several characters are listed are usually of two types: (1) those in which the first character will be difficult to find or determine with certainty in all specimens, and (2) those in which the first character does not always hold true, and in which it should be considered in relationship with the remainder of the characters given. The first type is much more common and the average specimen will have a certain character or the characters will be absent. In the second type, when all the sum of the characters in a cuplet is studied and totaled, the specimen should direct the worker to the next cuplet.

It should be remembered that we are still in a purely exploratory stage in our knowledge of the animals treated in this book. Many forms are known from but a single collection. In many cases there have been available to authors less than five complete specimens upon which to base conclusions. Consequently, there was no opportunity for the author to determine the normal range of variation within that species. Characters which may seem to separate two species if only a few specimens are studied may no longer be true when many specimens are examined. Conversely, characters which on the basis of a few specimens seem to be mere variations may, upon the examination of many individuals, prove to be decisive for the separation of a species. So it may very well be that the characters which have been used in the included keys will not hold when more specimens are available.

The worker using this book should, therefore, bear in mind that these keys are merely indices to the probabilities. Those probabilities should be weighed against agreement with the illustrations given and the known facts of distribution, both host and geographical, bearing in mind that those facts can change.

The design of the keys to families was so arranged as to allow the individual to quickly recognize those families that contain but a few described species thus building confidence in placing the specimen quickly in contrast to those families that contain large numbers of genera or species thus making it difficult to key to the species level. Another reason was to bring the reader into contact with the smaller families as he progresses through the keys and, so to speak, becomes familiar with these families.

It will be most helpful in the identification of a strange, unfamiliar specimen to merely look through the figures to pick out a similar form. I am constantly amazed at the ability of students who, having never seen a mite before in their life can, by looking at figures, place them in their correct family and never utilize the key. They find a familiar form and then go back and key the specimen, always with the figure in mind, and are more often right than wrong.

Identification by elimination is one of the fastest and most successful methods used by most people in attempting to determine a certain animal. For example, if the mite is listed as a parasite of a mouse and it has been taken from a mouse, you are able to narrow the possibilities to those mites associated with rodents. This technique will work for many free-living forms also as some are associated with distinct soil types such as pastures, forests or deserts. One learns to look for certain groups in certain types of habitats. However, do not be surprised if you find some forms in an area where it is not supposed to be. For example, it should be expected at times to find mites on a predator that is not its normal habitat due to the true host having been consumed by the predator. Also, free-living predacious mites are often found on a host animal such as a bird feeding on the feather mites that are parasites of the bird.

It should be remembered that keys, at best, are only artificial aids and that absolutely definitive identifications of closely related species must be made from comparison with authoritatively identified specimens or by careful checking with original or revised descriptions of the species.

Morphology of Mites and Ticks

A detailed treatment of the external morphology of each of the suborders is given. However, only those structures that tend to aid a worker in identifying mites and ticks will be discussed.

The gnathosoma, or head region, within the suborder Prostigmata tends to be fused. Many parts of this region may join the coxae of the pedipalps. The chelicerae may be styletlike with the endites of the pedipalpal limbs being fused. The basal portion of the chelicerae may fuse to form a stout region called the stylophore. The stylets are distal portions of the chelicerae. They curve upward attaching to the cheliceral plate at its ventral surface to form a closed pre-oral food channel into which the labrum projects dorsally. This is a major difference from the suborder Mesostigmata. In the genera *Tetranychus* and *Cheyletus* the pharynx is shortened and cup-shaped in structure. In most forms the digitus fixus is lost and is a reduced transparent scalelike structure above the mobile digit. Examples of this are seen in the Trombidiidae, Anystidae and Pseudocheylidae. The Bdellidae have the gnathosoma modified into a long conus buccalis. The long chelicerae lie in a trough which gives the Bdellidae (Fig. 2) a snouted weevil-like appearance. The palp is also modified, with the tarsus and femur being elongated and thin. The palp is bent in the middle.

The body of the pedipalpal segment, called the hypostome, bears the mouth at its anterior end. The hypostome (also referred to as the infracapitulum, subcapitulum or buccal cone in the Prostigmata) contains up to three pairs of adoral setae on the lateral lips.

The main feeding structures are the chelicerae. These are important in identifying many groups within the Prostigmata. The chelicera of the Prostigmata is three-segmented. A small basal section is considered to be homologous to the trochanter of the legs. The remaining segments that form the main body of the chelicera and the movable digit are considered to be a fusion of four segments, the femur, genu, tibia and tarsus.

Food habits within the Prostigmata range from parasitic to predacious, phytophagous, mycetophagous, necrophagous or coprophagous. Therefore, the chelicerae exhibit a variety of structural forms resulting from adaptation to a particular method of feeding. Most Prostigmata lack the ability to ingest solid food as do the Mesostigmata. Prostigmata such as *Balaustium florale* that feed on pollen of various plants use the chelicerae to pierce the pollen grain and suck out the liquid contents. The chelicera is normally chelate-dentate; however, in certain predacious species, like *Blattisocius*, they are modified and the teeth are reduced in size. In the phytophagous species a styliform chelicerae is found. In this type, the stylets are believed to be formed by the elongation of the movable digits. In the Tetranychidae, the cheliceral bases have fused to form an evertible stylophore. The pedipalps are movable appendages articulated to the gnathosoma and are normally sensory in function or modified for grasping food. In some Prostigmata one or more pedipalpal segments are armed with thick spines and spurs. The pedipalps may contain a dorsal claw on the tibia and the tarsus may be very short, thus forming a thumb-claw complex. This thumb-claw complex is utilized in separating out family groups in all keys to the Prostigmata. The pedipalps are also utilized in cleaning the chelicerae after feeding. The superfamilies Cheyletoidea and Trombidioidea are good examples of mites that have a palpal thumb-claw complex.

The family Cheyletidae shows the relationship between the peritremata and chelicera. The basal parts of the right and left chelicerae are fused together into a broad, integral cheliceral plate or stylophore. Inlaid into the surface skeleton of the stylophore are conspicuous, partially segmented peritremes. Segmented cheliceral peritremes of this type are also found in the family Syringophilidae. The protegmen is that part of the stylophore occurring in front of the peritremes. The configurations of the protegmen are utilized to describe differences in distinguishing genera within the family Cheyletidae.

In the superfamily Anystoidea the cheliceral bases are free, moving scissorlike over the gnathosoma. In contrast, members of the superfamily Cheyletoidea have cheliceral bases which are fused with the gnathosoma.

The idiosoma of members of the Prostigmata show a great diversity in the structure of the body plates. In the terrestrial forms, there is a tendency toward a reduction of the dorsal plating. However, in the aquatic forms these dorsal plates are well developed and utilized in their classification. The family Raphignathidae has a well-developed propodosomatal plate, a pygidial plate and two lateral occular plates.

The majority of the Prostigmata have a clear sejugal furrow which divides the idiosoma into a propodosoma and a hysterosoma. There may also be a third region, the opisthosoma, marked by the narrowing of the hysterosoma behind the fourth pair of legs. Ocelli are often present on the propodosoma. These may be on small plates or placed in the integument. There may be one or more depending on the species of the family. The idiosoma may be soft and striated. In the family Tetranychidae these striations are important in separating species.

The propodosoma and hysterosoma are independently sclerotized which allows flexibility between these two regions. This secondary development of dorsal plating occurs in both adults and larvae. In the Trombidiidae the dorsal plates are reduced to the propodosomatal shield in the larvae, together with a number of small platelets usually associated with setae as in the genus *Allothrombium*. In some adults of the superfamily Trombidioidea the propodosomatal plate is reduced to a crista metopica or a scutellata crista the details of which are of systematic importance, while the rest of the dorsal region is unplated. In the true aquatic mites there is extensive secondary development of dorsal plates.

Adults of the superfamily Tarsonemoidea have idiosoma covered by a number of overlapping horny plates. The propodosoma may overlap the gnathosoma, and in females often bears a pair of clavate pseudostigmatic organs which arise from cup-shaped pits.

The chaetotaxies of the dorsum within the Prostigmata may be very dense giving a hairy appearance, or may be sparsely distributed over the idiosoma. In forms with sparse setae the propodosoma may bear up to six pairs of setae, one or two pairs of which may be modified into pseudostigmatic organs. Many authors have used names, numbers or letters for these dorsal setae; however, no standard has been adopted by workers in this group of mites.

In the Tetranychidae (Fig. 1) the chaetotaxy of the body is important in establishing both tribal and generic categories. The setae are referred to by their position on the body of the mite, such as dorsal propodosomals, humerals, dorsocentral hysterosomals, dorsolateral hysterosomals, sacrals, clunals, and post anals. The problem of identification of a particular setae is a real one for the seasoned worker as well as the beginner. At the present there is no rigid terminology applied, due in part to lack of information regarding homologies of these structures throughout the Prostigmata.

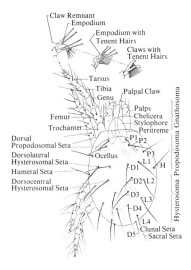

Figure 1

The coxae are a part of the body in most Prostigmata and the regions set off by the coxal apodemes are important in the classification of species. In the genus *Cheyletus* there are four separate coxal plates. In other groups the coxal plates of legs I and II and of III and IV tend to fuse forming two plates on either side of the body. This is seen in the genus *Cunaxa* in which the anterior pair meet and fuse at the mid-regions of the body. In many forms the ventral plating is reduced and includes only that area associated with the coxal surfaces. In the true aquatic mites there is a development towards four dorsal and four ventral plates.

The genital opening is normally located on the venter of the opisthosoma. Genital discs are common in many families. The opening may be transverse as in the Eriophyidae and Tetranychidae. The genital area may be covered by a shield as in the Trombidiidae and Johnstonianidae. The anal opening is terminal or subterminal.

The legs of Prostigmata are made up of the following segments: anterior to posterior, coxa, trochanter, femur, genu, tibia, tarsus and apotele. The femora may be divided in such families as the Trombidiidae and Erythraeidae. Associated with the femur of leg IV of the Anystidae there is a true subdivision of the femora in adults. In the Prostigmata the coxae are incorporated into the ventral surface. Some leg segments are absent in the parasitic Demodicidae. In some species of Tarsonemidae the genu and tibia of legs IV are fused.

The number of legs varies from zero to the full complement of eight. The Eriophyidae have four legs or two anterior pair. Males of the family Podapolipidae have from three to four pairs of legs while females have from zero to three pairs. The solenidia are normally bacilliform; however, they may be setiform and difficult to distinguish from true setae. These special setae are very important in the separation of species in many of the families of the Prostigmata. In the Tetranychidae the sensory setae are found not to be constant in number in contrast to other families within the Prostigmata. In the genus *Tuckerella* the solenidion are associated with tenent hairs, empodium and claws.

The tarsus of all legs may contain many

structures that are important in classification of members of the Prostigmata. The tarsus may contain true claws, empodium, tennent hairs and setae. The empodium may be modified into a sucker as in members of the Heterocheylidae, elongate with hairs in the Ereynetidae, bladderlike as in Pyemotidae, or clawlike featherclaws as in Tetranychidae. Tenent hairs are commonly found within families of the Prostigmata.

Leg IV in male Tarsonemids may be shortened and functions in clasping the female rather than in walking.

The true aquatic Prostigmata have legs which are fringed with elongate setae (hairs) for swimming. The tarsus of leg IV lacks the claw and contains several long setae in the Scutacaridae. In the Harpyrhynchidae legs III and IV are without claws and with long setae in contrast to the Cloacaridae in which all legs are short, stubby and contain spinelike setae. In the Teneriffiidae the tarsal claws I and II are strongly rayed. The members of the family Myobiidae have their first pair of legs greatly modified in order to clasp the hair of their mammalian host. The Psorergatidae possess a large ventral spur on the femora of each leg.

The larvae of all Prostigmata have three pairs of legs. The larvae of the Trombiculidae (chiggers) are important in that most of the family classification is based on the larvae.

Internal structures include a stomach that ends blindly. The posterior body opening, called a uropore, is the opening of a single median excretory canal. The esophagus is a narrow tube with a thin chitinous lining. There may be as many as ten caeca in *Trombidium*. In the genus *Cheyletus* the stomach is small and there is a pair of large posterior caeca. There may be as many as six paired and a single unpaired gland asociated with the mouth. These glands produce substances that aid in digestion. In *Cheyletus* the gland

secretion causes extensive histolysis of prey tissues.

The hind part of the gut functions as an excretory mechanism. However, in *Tetranychus telarius* there is a communication between the single median excretory canal and the anterior dorsal wall of the stomach. This can be correlated with the habit of sap feeding. The food intake, in the case of sap, contains relatively large proportions of water, which would be difficult to dispose of from the usual blind gut of the Prostigmata. In the Prostigmata the excretory tubules have lost their function, which has been taken over by the posterior stomach region.

The central nervous system is a solid mass of nervous tissue pierced by the esophagus, this mass of tissue is divided into a part dorsal to the esophagus and a ventral lobe. The reproductive system in the male may consist of testes which remain separate as in the *Tetranychus* or the testes and its duct may be single median structures as in the Tarsonemidae. Male accessory glands in the Prostigmata are numerous and complex. An intromittent organ is widely developed in the Prostigmata. The penis may be heavily sclerotized as in some species of the genus *Cheyletus*.

Females contain a single median ovary in the Anystidae and Bdellidae, and paired oviducts in the Rhagidiidae and Cheyletidae. Accessory organs in the form of glands, receptacula seminis and bursae copulatrices may be present.

The Mesostigmata (Fig. 3) has the body divided into two regions with most members flattened dorsoventrally. The gnathosoma, or head region, is attached anteroventrally to the idiosoma, or body region. The idiosoma of mites corresponds to the thorax, abdomen, and a portion of an insects head.

The gnathosoma is tubular, formed by the appendages of the first and second segments, the chelicerae and pedipalps. The upper plate of the gnathosomal tube is the

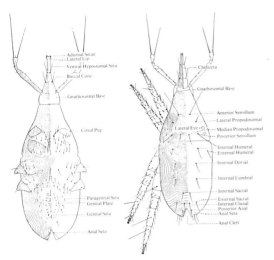

Figure 2

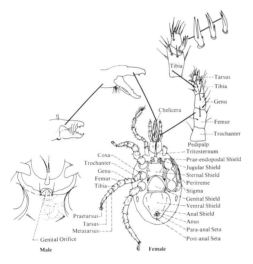

Figure 3

tectum or epistome with the lateral sides being the fused coxae of the pedipalps. The lower section or plate of the gnathosoma is the subcapitulum. The hypostome is a combination of the internal mala, coniculus, anterior coxal endites of the subcapitulum and other structures of the gnathosoma.

The basis capituli is the sclerotized portion of the gnathosoma enclosing the chelicerae, or mouthparts. In many Mesostigmata the epistome is projected anteriorly to form a roof called the rostrum. The anterior margin of the epistome may be smooth, enticulate or formed into different simple or complex processes. The basis capituli has its lateral and ventrolateral walls formed by the coxae of the pedipalps and these are separated ventrally by the hypostome. The hypostome has a ventral or capitular groove which has rows of denticles. The capitular setae are found on either side of the capitular groove. The important and conspicuous corniculi and internal malae are associated with the rostrum. The rostrum contains three pairs of setae which are the rostrals and the paired interior and posterior rostrals. These are important taxonomically in identifying Mesostigmata.

The chelicerae are preoral appendages essentially used as feeding organs. In the majority of the free-living forms the chelicerae are three-segmented. There is a short basal segment and a larger distal segment of the shaft which terminates in what is called a fixed digit. The movable digit which would be the third segment is ventrolateral to the distal segment. In the nonspecialized free-living species which make up the greater part of the Mesostigmata the chelicerae are referred to as chelate with the opposing edges of the digits provided with distorted teeth. Also on the face of the fixed digit there is a dorsal seta, dorsal pore, a large lyriform pore and one or more ventral setae. This fixed digit may also have a specialized seta called the pilus dentilis or dentarius. This seta is used in many groups as an important structure in separating closely related species. Its variation in size and shape is the major reason for its usefulness in taxonomy. The spermatophoral process found in males of some species is a part of the movable digit and arises from the proximal half of the digit. It may be fused distally with the digit or free of the digit completely. The structure is an accessory copulatory organ used to transfer the spermatophores from the genital opening of the male to that of the female. The digits of the chelicerae may undergo modification in their structure due to the wide range of feeding requirements of the members of this group. In the genus *Dermanyssus*, which feeds on birds, both digits have undergone specialization and developed into a styletlike organ for piercing the skin of their host. In some species the digits are devoid of teeth or one digit is shorter than the other. These differences aid in separating various species and higher taxonomic categories.

In a study of the feeding organs, Grandjean 1947, considered the primitive limb of his group Actinochaeta to be made up of seven segments. The terminal segment, the apotele, would form the ambulacrum of the walking appendage. Therefore, the movable digit of the chelicera would be homologous with the apotele. Carrying this on to completion, the fixed digit would therefore be regarded as being derived from birefringent setae on the penultimate segment. Since members of this group contain chelicera that are three-segmented, we would expect to find a small basal section, homologous with the trochanter of other appendages, a large compound segment which constitutes the main body of the chelicera and is considered to have been a product of the fusion of four segments, the femur, genu, tibia, and tarsus. The short movable digit then completes the third segment of the chelicera.

Members within the Mesostigmata show diverse methods of feeding habits, therefore it is not unexpected to find structural modifications particularly with regard to the chelicerae. However, within the Mesostigmata and mites in general the chelicera is usually chelate-dentate.

The pedipalps are divided into five segments and resemble the walking legs in structure. They are attached to the anterolateral angles of the basis capituli by the trochanter. From the trochanter the remaining segments are: femur, genu, tibia and tarsus. This is similar to the segments of an insects leg except for the genu which is not found in insects. Some authors refer to the pedipalps as having six segments and include the palpal apotele as a segment. In this case there would be six segments of the palp. The palpal apotele is almost invariably reduced and resembles a modified setae. The setae, referred to in most works as the chaetotaxy of the trochanter, femur, and genu are utilized as a source of valuable taxonomic characters. In the adults of those species that have not undergone some type of specialized host adaptation the trochanter, femur and genu would have two, five, and six setae respectively. The structure of the palptarsal setae in most of the species that are associated with animals as parasites carries ventrally on its internal basal angle a conspicuous pronged seta. The development of this setae has been utilized as an indicator of a parasitic habit. The setae may be two, three or four pronged. If two pronged, it is referred to as two tined and usually restricted to parasitic forms. However, there are exceptions to this rule. In any case the beginner would be ahead to check the keys to parasitic forms first when encountering the two pronged condition on a specimen. In species where this setae is three or four pronged the prongs may be subequal or markedly different in length. In the family Veigaiidae, which is free living in habit, this setae is associated with a large leaf-like structure. The number of setae on the pedipalpal segments in Mesostigmata is a useful character in separating immature stages of certain species.

The idiosoma is a composite structure and considered by some workers to be the body of a mite. In the Mesostigmata the idiosoma is generally round or oval in shape; however, there may be exceptions to this due to a specialized type of living habit. Species associated with a particular host animal may develop according to its position on that host, thus the idiosoma would assume a shape determined by this position. The dorsal surface of Mesostigmata mites as well as the remaining idiosoma shows no definite evidence of primary segmentation. In most forms there exists both dorsal and ventral shields or plates. These shields in the adult stages may completely or partially cover the idiosoma. The dorsal shield may be entire, deeply divided, or separated into two or more divisions. In the genera *Veigaia* and *Arctoseius* the dorsal shield is divided laterally. The idiosoma is divided by most workers into several regions based on the position of the four pairs of legs which are found on the propodosoma, and hysterosoma. The propodosoma bears the first two pairs of legs and the hysterosoma the last two pairs of legs. The name prodosoma is utilized by some workers to include the body region that contains the four pairs of legs and does not include the posterior region behind the fourth pair of legs. This posterior region is called the opisthosoma and is a part of the hysterosoma (metapodosoma by authors).

There is a circumcapitular suture that separates the gnathsoma (head), from the body region (idiosoma). The groove or constriction found in many of the other arachnids, such as the spiders, behind the last pair of legs is absent. The dorsal shield or shields may, depending on the species, be reticulated, beset with small or large punctuation, or completely without such modifications and be smooth. The interscutal membrane which forms the border of the shields and the shields themselves is often beset with setae utilized in identification of certain genera and species. The anterior dorsal shield in many forms is enlarged and may cover all the leg-bearing segments. However, in the endoparasites this shield can be greatly reduced or absent altogether such as in members of the genus *Pneumonyssus* which are parasites of the lungs of certain monkeys. The chaetotaxy

of the dorsal shield is used to distinguish certain families within the Mesostigmata. This is particularly so when keying out the families Phytoseiidae and Ascidae.

The chaetotaxies for the dorsum of adult Mesostigmata utilize in most cases letter symbols to denote the particular row of setae. For example, the dorsal row according to the system of Evans, 1957, would be a capital D, the median row a capital M, the lateral row a capital L, and the marginal row by a combination of a capital M and a small g (Mg). The specific setae would be numbered beginning with the most anterior setae as D, and progress posteriorly to the last setae of that row (D1, D2, D3, ... DN). However, not all workers have followed this system. Hirshmann 1951, devised a system in which other letters are utilized.

The dorsal shields in the Mesostigmata, when divided, are named depending on the region of the idiosoma they cover. For example, if the shield covers the area of the idiosoma from the circumcapitula suture to an area between the second and third pair of legs it is named propodosomal shield (podonotal of some authors). A dorsal shield covering the area posterior to the region described above is from the area of the third pair of legs to the area just below the fourth pair of legs and would be named the hysterosomal shield. The beginner can become confused by the use of the terms opisthosomal shields and pygidial shields. This stems in part from the fact that a knowledge of the larva and protonymph stages of the Mesostigmata is required. The dorsum of these stages is usually provided with an anterior podonotal shield, one or more pairs of mesonotal scutellae and a pygidial shield. The formation of the opisthosomal shield (hysterosomal shield) is by the fusion of the mesonotal scutellae. In those forms where the fusion has encompassed the pygidial shield the terms opisthosomal or hysterosomal are used. However, in some adults within the Mesostigmata there is a pygidial shield present and is so named.

The ventral region of members of the Mesostigmata has some of the most important structures used in the identification of this group of mites. In most free-living mites the ventral region is covered by a number of shields (plates by some authors) that are well defined. The coxae are the first structures that a beginner should locate, as they can be the landmarks from which to trace the plating of the remainder of the ventral region. The first important structure to locate on the venter is the sternite of the segment bearing the first pair of walking legs, the tritosternum. This can be found by inspecting the region between the first pair of coxae (coxae I). In this area the base of the tritosternum will be located and its extremities which consist of a pair of smooth pilose lacinae extend to the ventral region of the gnathosoma (Fig. 3). The tritosternum may be absent or only the base present in some parasitic forms. It should be emphasized at this point that specimens of Mesostigmata mites that are mounted with the dorsal side up and not completely cleared of their internal contents, may have many of the ventral plates not visible, the tritosternum being one of them. Even material mounted with the ventor facing up and not completely cleared of body material may give the worker difficulty in finding the tritosternum. This structure is extremely important in keying out the families contained within the Mesostigmata as the presence or absence of the tritosternum separates some families such as Dasyponyssidae and Spinturnicidae which are parasitic on edentates and bats respectively. The width of the base is utilized by some workers to separate the superfamilies Protodinychoidea and Trachytoidea. Posterior to the base of the tritosternum one or more pairs of small plates will be found. These are called the presternal shields (prae-endopodal shields, praeendopodal plate by some

authors). It should be noted by the worker that in many forms the sternal shield may become divided and a pair of shields bearing the first pair of sternal setae and called the jugular shields may be detached and occupy the same region as the presternal shields. The problem for the beginner is to determine which are jugular and which are presternal shields. This can be done as only the jugular shields bear setae similar in structure to those found on the sternal shield. The sternal shield normally is posterior to the base of the tritosternum with its anterior end just posterior to coxae I and in the female extends to the region between coxae III and IV depending on the particular species being identified. In the male of many Mesostigmata the whole ventral plate region becomes fused.

Posterior to the sternal shield in the female is the epigynial shield (genital shield by some authors). Between the sternal shield and the epigynial shield is the genital orifice and in the Mesostigmata it is usually a transverse slit. The epigynial shield may extend anteriorly over the genital orifice and form a groove along which eggs may travel during oviposition. The structure of the epigynial shield has been of major importance in constructing keys to separate not only various genera and species but also to establish the higher categories such as supercohorts, cohorts, superfamilies and families within the Mesostigmata. The epigynial shield may be well-developed and well-sclerotized in the free-living forms or weakly sclerotized in many of the parasitic species. The setae and pores associated with the epigynial shield are referred to as genital setae and pores. These associated with the shield either lie on the shelf itself or lateral to it on the interscutal membrane. These setae and the epigynial shield are important taxonomic structures as they are the first structures that a worker encounters in most keys used to separate the various taxa within the Mesostigmata. The number of setae present on the epigynial shield is also important in distinguishing higher taxa. One difficulty a beginning worker will encounter is that some genera such as *Dipolyaspis* and *Protodinychus* are devoid of genital setae. However, each have a pair of setae on the epigynial shield. Most workers who have worked with these groups are aware that these setae are the metasternal setae and are distinguished from true genital setae in that they are accompanied by two pairs of pores.

The ventral shield is located between the posterior margin of the epigynial shield and the anal shield. The ventral shield may in some forms be fused with the epigynial shield and called a geniti-ventral shield. In some species the male has a holoventral shield which is a fusion of the sternal, epigynial, ventral and anal shields. Several other small shields may occur posterior to coxae IV; however, only the metapodal shield is of any concern for the worker attempting to identify Mesostigmata. This pair of shields is located on either side of the ventral shield and may fuse with this shield.

The setae present on the ventral surface of the idiosoma are utilized to distinguish the different taxa within the Mesostigmata. The most important setae are usually associated with the shields, therefore a worker must not only locate the specific shield but must in many cases determine the correct number of setae present on the shield.

All of the structures that have been discussed previously for the venter have been with shields that are located between the coxae of the four pairs of legs or the inner region of the venter.

The outer or lateral area also contains important structures. The paired stigmata are situated ventrolaterally in the region of coxae III and IV. Associated with each pair is a long sclerotized tubelike structure, the peritreme. In some parasitic forms the peritreme is reduced or absent. The peritreme extends anteriorly from the stigma to beyond coxae I. The peritreme is an excellent structure for distinguishing the Mesostigmata as a group. The peritreme in some species passes from the venter and has its most anterior end on the dorsal surface of the lateral region of the idiosoma. This is due to the enlargement of the coxae or the dorsolateral compression of the idiosoma. Surrounding the peritreme is the peritrematal shield and in many forms it may be fused with the parapodal (expodal by some authors), and ventri-anal shield to form a groovelike area in which the peritreme is embedded. In some genera the peritremal shield may, when fused with other shields, cover a large portion of this opistosomal region.

All stages of Mesostigmata have legs that are comprised primitively of seven segments. The division of the legs is named similar to that used for insects, namely, coxa, trochanter, femur, genu, tibia, tarsus, and apotele (ambulacrum). The femur may in some species be divided into basifemur and telofemur, as well as the tarsus which may be divided into a metatarsus and tarsus. Like the femur and the tarsus even secondary division may occur on the trochanter. The termination of the tarsus, the apotele, usually contains two claws and a padlike pulvillus (empodium). The claws may be strongly developed or in some species on the first pair of legs be completely absent. The setae of the leg segments are utilized by many workers as a means of distinguishing different species. Because of this use chaetotactic formulae have been made by various authors. Legs I are usually sensory in function. Also all legs in some groups may be withdrawn into deep cavities in the idiosoma. In the Mesostigmata, like the Metastigmata (ticks), the larval stage is readily distinguished from the other life stages by having only three pairs of legs (leg I-III). In the male of certain forms leg II and sometimes leg IV may contain very stout spurs that are utilized in clasping the female during copulation thus showing some sexual dimorphism.

There are within the Mesostigmata four morphologically distinct post-embryonic life stages: larvae, protonymph, deutonymph, and adult (male and female). The beginning worker will find that it is the protonymphs and deutonymphs that will give him the most trouble, as many times in obtaining specimens only these two stages are found and they will not have the adult structures and cannot be keyed out to their respective species group. Therefore, it will save a worker considerable amount of time and frustration to be able to separate these immature forms from adults. A worker should sort out the immature stages and use only adult forms for identification to the different taxa. This is not to minimize the importance of these stages as the immature stages are in need of much work. However, this should be done by competent specialists. In the free-living species of the Mesostigmata the proto- and deutonymph may be separated by the chaetotaxy of the trochanter and the femur of the pedipalp.

Internal structures include the digestive system beginning with the mouth which is connected to the stomach (ventriculus) by the esophagus and the salivary glands. The stomach of the Mesostigmata is small. The esophagus enters the stomach on the antero-ventral surface and projects slightly into its cavity. This arrangement prevents regurgitation of the stomach contents. There is a common opening of the Malphigian tubules. Lateral and dorsal caeca are present which extend anteriorly and posteriorly and are quite large. A smaller pair of caeca arise from the posterior wall of the stomach ventrally extending to the posterior region. A short intestine which is called the foregut leads into a large vesicular region, the midgut. The Malphigian tubes open into the hind gut. This lateral structure may be ovoid or spherical and separated from the intestine by well-marked constrictions in many species. The paired excretory ducts open into the midgut and communicate by rectal valves with the anus.

The vascular system consists of a circulatory organ termed a heart and located in the anterior of the opisthosoma. It has a single chamber and one pair of ostia. The heart has a cephalic aorta but the aorta caudalis is absent. Also the aorta does not enlarge into a periganglionic sinus which is characteristic of the Metastigmata.

The nervous system consists of a supra-esophageal ganglion which may be divided into a paired cerebral and cheliceral ganglia. This region of the brain is penetrated by the esophagus. The paired cerebral ganglion innervates the dorsal pharyngeal muscles, the labrum-epipharynx and the ventral setae. The cheliceral ganglia serve the chelicerae. There is a pedipalpal ganglia which supplies nerves to the pedipalps, pharyngeal musculature and the rostral setae.

The subesophageal ganglion has four pairs of ganglion which supply nerves to the legs and associated musculature.

The female genital organs include the unpaired ovary, the oviducts which are paired structures and a median uterus. In the Mesostigmata there are two types of vaginal invagination: the parasitid type which contains the uterus and vagina; and in most free-living forms which have a single structure, the uterus. In the males a penis is entirely lacking, therefore a spermatophore is placed in the female genital opening by means of the chelicerae. The testes are tubular, a pair of vasa deferentia, a seminal vesicle, an ejaculatory duct and accessory glands are present.

In ticks the cephalothorax and abdomen are fused into an oval or elliptical body, flattened dorsoventrally, bearing four pairs of six-jointed legs and a false head, the capitulum.

In the hard ticks (Family Ixodidae)

(Fig. 4) the capitulum projects from the anterior end of the body, while in the soft ticks (Family Argasidae) (Fig. 5) it extends from beneath the anterior end. This is the movable anterior portion of the body including the mouthparts. In the hard ticks there is a basal plate, the basis capituli, the shape of which is of taxonomic importance. The mouthparts consist of the hypostome, chelicerae, and pedipalps. The hypostome is called the median ventral mouthpart, is immovably attached to the basis capituli, and usually covered with rows of recurved filelike "teeth" or denticles used to anchor the tick to its host. The dentition is indicated by numerals on either side of a line. Thus 3/3 means that there are three longitudinal files on each half of the hypostome. The denticles or recurved teeth enable the tick to attach itself to its host so firmly that it is difficult to remove it from the host without tearing the capitulum from the body of the tick. The tick itself, however, withdraws its mouthparts quickly by slipping the hoodlike portions of the capitulum over the relaxed mouthparts and with a quick jerk, drops off the body of its host. The lateral paired chelicerae are long, cylindrical, chitinous shafts enclosed in mandibular sheaths. They are used for piercing the skin and are inserted when the tick feeds. Each chelicera terminates in a small toothed, retractile, chitinous digit of two movable parts. They are used as a cutting organ to permit the insertion of the hypostome. The chelicerae are so efficient that the host is completely unaware the tick has cut the skin. The paired flaplike pedipalps are of various shape and do not penetrate the tissues but serve as supports for the other mouthparts and in some species as sense organs. This structure is generally referred to as the palpi. Each palpus is an articulated appendage attached to the basis capituli lying parallel with and sometimes enclosing the hypostome and chelicerae. Four segments were originally present in all ticks,

but segment one may be vestigial or absent in some ticks. The articulation between the basis capituli and segment one, if present at all, is represented only by a sclerotized spot, or plate, on the inner side which may be very indefinite, a distinct inner edge, or a retrograde extension, which, because of its shape may be called a "ligula." The palpi in some genera unlike most ticks do not have a fixed length as may be observed in occasional specimens which have one palpus distinctly longer than the other. In *Rhipicephalus*, segment one is present and movably attached to segment two. On the venter there is a large triangular plate on the inner side. In the species *Haemaphysalis leporispalustris* (Packard) palpal segment one is absent and the articulation between two and three is fused. In all other ticks the palpus is made up of four segments designated by numerals from 1 proximal to 4 distal.

The basis capituli (often abbreviated to "basis") contains some important taxonomic structures used in classifying hard ticks. It is movably attached to the anterior part of the scutum of hard ticks. In many species it lies between the scapulae (the anterior angles of the scutum). In soft ticks it is sometimes referred to as the "basal ring" and is not seen from a dorsal view. In the hard ticks the basis capituli contains projections on both the dorsal and ventral regions. The auricula (sometimes called "ears" or "horns") are paired ventral extensions at the sides of the basis capituli. They are often absent in some species. The cornua are paired caudad projections extending from the posterior lateral dorsal angles of the basis capitulum. The length of each cornu is important in distinguishing species of the genus *Dermacentor*. The overall shape of the basis capituli may serve to distinguish between different genera of some hard ticks. In the genus *Rhipicephalus* the hexagonal shape of the basis capituli may serve to readily separate it from other genera

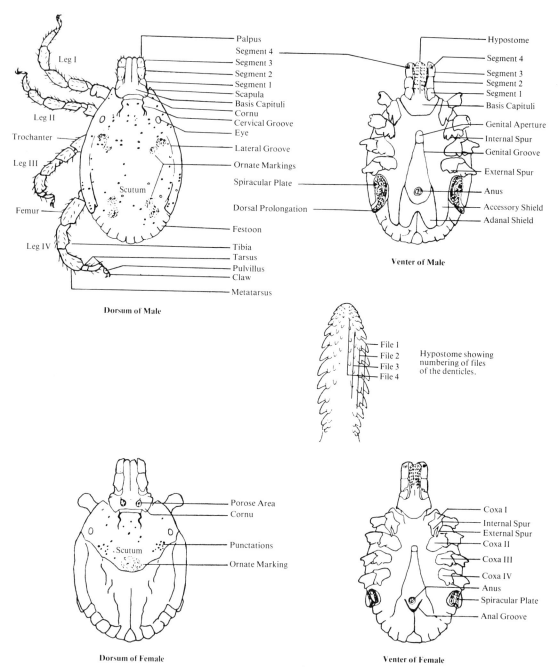

Leg I

Leg II

Trochanter

Leg III

Femur

Leg IV

Scutum

Dorsum of Male

Palpus
Segment 4
Segment 3
Segment 2
Segment 1
Scapula
Basis Capituli
Cornu
Cervical Groove
Eye
Lateral Groove
Ornate Markings
Spiracular Plate
Dorsal Prolongation
Festoon
Tibia
Tarsus
Pulvillus
Claw
Metatarsus

Hypostome
Segment 4
Segment 3
Segment 2
Segment 1
Basis Capituli
Genital Aperture
Internal Spur
Genital Groove
External Spur
Anus
Accessory Shield
Adanal Shield

Venter of Male

File 1
File 2
File 3
File 4

Hypostome showing
numbering of files
of the denticles.

Scutum

Porose Area
Cornu

Punctations
Ornate Marking

Dorsum of Female

Coxa I
Internal Spur
External Spur
Coxa II
Coxa III
Coxa IV
Anus
Spiracular Plate
Anal Groove

Venter of Female

Figure 4

with a distinctly different shaped basis capituli.

The body color of ticks is usually reddish or mahogany brown, some species show

a distinctive ornate color pattern. The hard
ticks have a dorsal shield, the scutum, which
almost covers the dorsal surface in the male
and about one-third or one-half of the ante-

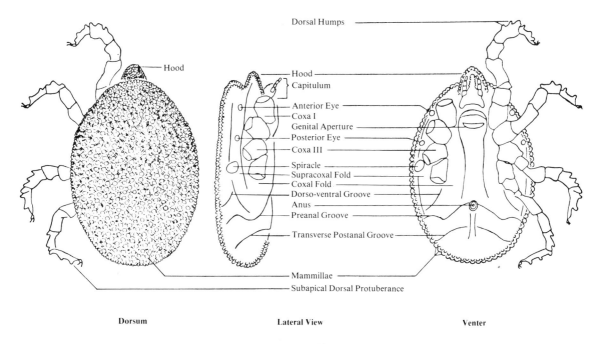

Dorsal Humps

Hood

Hood
Capitulum

Anterior Eye
Coxa I
Genital Aperture
Posterior Eye
Coxa III

Spiracle
Supracoxal Fold
Coxal Fold
Dorso-ventral Groove
Anus
Preanal Groove

Transverse Postanal Groove

Mammillae
Subapical Dorsal Protuberance

Dorsum   Lateral View   Venter

Figure 5

rior dorsal surface in the female, nymphs and larvae. The term pseudoscutum is applied to the anterior portion of the scutum of the male which corresponds with the scutum of the female. In many species it is not clearly marked, but may be somewhat differentiated by faint surface markings or punctations of a different size. A scutum is absent in all soft ticks as well as ornate coloration. The eyes, when present in hard ticks, are generally on or near the lateral margin of the scutum, usually small, simple, shining organs. They are not found on the basis capituli as might be expected. They are "marginal" if situated on the edge of the scutum, "not marginal" if placed a short distance away from the edge of the scutum. The eyes are often present on nymphs and larvae of hard ticks. In the soft ticks, if present, the eyes are on the lateral or supracoxal folds, this would correspond to the side of the tick.

In hard ticks the posterior margin of the body is often divided into rectangular areas, called festoons. These are separated by grooves along the posterior submarginal area of the dorsum in both sexes. In *Haemaphysalis* they are present on nymphs and larvae as well as adults. The folds or constant ridges of the integument found on the venter in soft ticks are not called festoons.

Ticks have four pairs of six-jointed legs, usually terminating in claws, arising from shieldlike plates, called coxal plates, on the ventral surface. The coxal glands between the first and second coxae secrete a fluid during feeding and copulation. An olfactory organ, Haller's organ, is on the dorsal surface of the tarsus of the first pair of legs. This structure contains sensilla that are stimulated by scents and humidity. It is the use of this organ that enables ticks to sense the approach of a host animal. The tick positions itself on grass or other suitable vegetation and at intervals waves the first pair of legs in the air. This reaction is brought about by mechanical disturbance or shading, such as would be produced by the approach of a host animal. The same reaction of extending the legs can be brought about by a rise in temperature accompanied by the smell of animal hair. It has been determined that these olfactory sense organs are the peglike chemoreceptors in the capsule of the Haller's organ, the temperature receptors are short setae on the dorsal surfaces of the legs. It is believed that the light sensitivity is a general cutaneous property. Haller's organ is one of the anatomical structures used in classification as separating the ticks from their close relatives. The number of segments in the typical leg of a hard tick is six: coxa, trochanter, femur, genu, tibia and tarsus. The coxa in many species of ticks contain spurs. The spurs are of great significance in the classification of certain genera. These structures are posterior projections on the posterior side of the coxae. Spurs may be large or small, pointed or rounded, present or absent. When there are two spurs on coxa I and they are long, they may be referred to as bifid. Other types of spurs are trochantal spurs which are retrograde projections near the distal end of the trochanter. Both dorsal and ventral spurs may be present in a species. In the soft ticks the leg segments are labeled differently by some workers. The number is still six but are called: coxa, trochanter, femur, tibia, metatarsus, and tarsus. Soft ticks have protrusions called dorsal humps. The humps or elevations on the dorsal walls of the leg segments along with the subapical dorsal protuberance are important in classifying members of the family Argasidae.

The spiracles are located on chitinous plates directly behind the third pair of coxae in the soft ticks and behind the fourth pair of coxae in the hard ticks. The chitinous spiracular plates of hard ticks of the genus *Dermacentor* are utilized to separate species. The shape, completely oval or oval with a dorsal

prolongation and the size of the goblets are important structural characters.

The anus is near the posterior margin, but the genital opening is well forward on the mid-ventral line. The genital grooves extend from the genital opening to the posterior margin of the body. There is usually an anal groove behind the anus which may connect with the genital grooves. An exception to this rule is found in the genus *Ixodes* of the hard ticks where it forms an arc in front of the anus and extends to the posterior margins without contact with the genital grooves. These grooves are not apparent in the soft ticks. On some genera of hard ticks, ventral plates are present. In males of some genera, there are paired ventral plates bordering on the median and anal plates called adanal plates. In the genera *Boophilus* and *Rhipicephalus* males have adanal shields, paired ventral shields near the anus, and accessory shields, paired shields outside the adanal shields.

Internally the digestive system is divided into foregut, midgut, and hindgut. The foregut includes the tubular buccal cavity; the pumplike chitinous muscular pharynx; the thin-walled S-shaped esophagus, and the paired salivary glands. The midgut consists of a large, four-lobed, thin-walled stomach that serves for food storage. The hindgut is made up of a narrow rectum and a large vesicular rectal sac which receives the two long, convoluted Malpighian tubes.

The vascular system consists of a heart that is dorsal to the stomach, a cephalic aorta that runs forward to the periganglionic sinus, and four arteries leading from this sinus to the legs. Contraction of the dorsoventral muscles causes the blood to return to the heart through the lacunar spaces. A highly developed tracheal respiratory system connects the spiracles with the organs.

The nervous system consists of a large,

anterior, midline ganglion with numerous nerve trunks.

The female genital organs include the single ovary that opens into two coiled oviducts, a uterus, vagina, accessary vaginal glands, and Gene's organ which supplies secretions during oviposition. The male has lateral paired testes, a pair of vasa deferentia, a seminal vesicle, an ejaculatory duct, and accessory glands.

Species within the Astigmata (Fig. 6) may range in their feeding habits as saprophytic, fungivoric, predaceous, phytophagous or parasitic. Within the Astigmata are the mites most commonly found feeding on stored food products. The common species, *Acarus*

siro, which tends to be restricted to grain which has absorbed moisture, can destroy grain completely. *Glycyphagus domesticus* and *Tyrophagus putrescentiae,* which feed mainly, if not exclusively, on microorganisms growing on stored material, are common inhabitants of stored food products. Species such as *Carpoglyphus lactis* tend to be more restrictive in their feeding habit and occur on dried fruits, jams, and cheese. Both *A. siro* and *Gohieria fusca* are common in bagged flour. Members of the genus *Rhizoglyphus* feed on decaying plant material and fungi, while members of the genus *Histiogaster* are common on fungus-infested tree bark.

The Astigmata are not regarded as soil

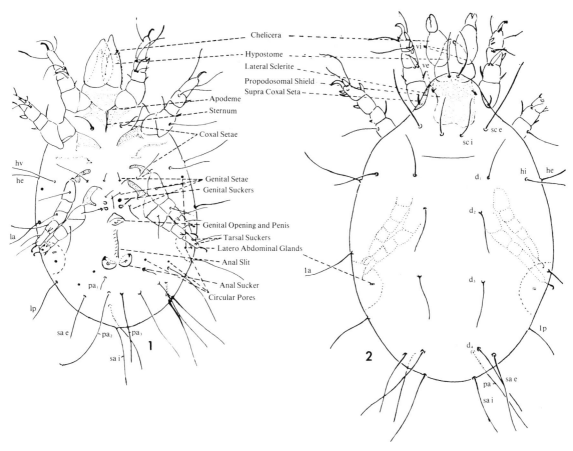

Figure 6

inhabiting. However, certain groups, such as members of the genus *Rhizoglyphus* (especially *R. echinopus*) are commonly found in the soil. In contrast, this group of mites contains the majority of epizoic forms that feed on the skin secretions of birds and mammals or in many cases the scales either from the feathers of the birds or the hair scales of mammals. This group has normally-developed chelate chelicerae which are not highly specialized as in the true parasitic forms. Examples of this group are the feather mites of the superfamily Analgoidea and the fur mites of the superfamily Listrophoroidea.

The true parasitic Astigmata include a large number of forms that parasitize warm-blooded animals. The psoralgids and psoropids are found feeding on the skin surface of mammals. The genus *Chorioptes*, with mouthparts specialized in abrading the skin surface to obtain the liquid secretion of its host, causes skin irritation. *Psoroptes* species have mouthparts specialized to pierce the epidermis and penetrate the dermal capillaries of the host thus causing a condition commonly referred to as mange on many domestic mammals. The species of the genus *Otodectes* are the common ear mites. The other groups of Astigmata all have their mouth parts adapted but in a less specialized manner than those described as true parasites.

The gnathosoma is the feeding apparatus partially retractile, very mobile and normally held at an angle to the idiosoma so that the apex of the chelicerae can come into contact with food materials. There is a pair of anteroventral expansions, the external malae, formed mainly by the fused coxae of the pedipalps. The roof of the gnathosoma within the Astigmata is formed from the dorsal fusing of the pedipalpal coxae. Many of the species have undergone considerable modification of the gnathosomal region to adapt themselves to a particular type of feeding habit. Members of the family Anoetidae, which feed on mi-

nute particles of food material in a liquid medium, have the under structure expanded to meet above the chelicerae. In the families that are associated with animals such as mammals, adaptation of the coxal region of the pedipalps may be utilized in clasping the hair of its host. The remaining segments of the pedipalps are sensory appendages or modified for grasping food. The last two segments are normally freely movable and very specialized. The terminal segment will contain setae called solenidia. In the genus *Histiostoma* the palp is flattened and used to push liquid food material into the mouth.

The idiosoma in Astigmata mites usually contains a propodosomal shield and occasionally a sclerotization of the hysterosomal area called the hysterosomal shield. The propodosomal shield is well developed in the Analgoidea (the feather and skin parasites of birds) and may be extended by the incorporation of setal bases and a merging of the hysterosomal shield to cover the entire dorsum with only small membranous areas separating the body regions. In the free-living groups the dorsal propodosomal shield may be present, reduced and in some, completely absent. The dorsal shields may consist of a single propodosomal shield alone or in combination with an opisthonotal shield or a pygidial shield. The latter two may be fused together and split longitudinally in which case the two are called opisthonotal shields.

The idiosoma may be fusiform, convex with the integument smooth with minute setae as in members of the family Fusacaridae, or oval with the integument strongly or lightly pigmented, smooth, wrinkled, or with fine projections. There may be a transverse sejugal furrow between propodosoma and hysterosomal regions as in the members of the family Saproglyphidae. The integumentary striae may form patterns resembling scales on the opisthosoma as in members of the families Yunkeracaridae and Listrophoridae, or many line-like striae as in the subfamily Labido-carpini of the Chirodiscidae.

The dorsal setae are very important in identifying species within the Astigmata. Setae to note especially are the vertical internal setae (v-i) which arise in the midline on the anterior dorsal edge of the propodosoma and project out over the gnathosoma. There is also an outer pair, the external verticals (v-e) which are behind the internal vertical setae and on the lateral margins of the propodosomal region. The external and internal scapulars (sc-e, sc-i) are across the posterior dorsal region of the propodosoma. The anterior lateral edge of the hysterosoma normally between legs II and III contains from one to three pairs of humeral setae called internal (h-i), external (h-e), and ventral (h-v) humerals. There are normally four pairs of dorsal setae (d_1 to d_4) arranged in two longitudinal rows on the hysterosoma. Lateral setae (L_1 to L_3) are found on the sides of the body. Two pairs of sacrals called the internal (sa-i) and external (sa-e) arise on the posterior edge of the body. There also may be a postanal or anal setae. Ventrally the setae are either associated with the coxae, genital region or anal complex. The coxal setae (cx_{1-4}) would be associated with the corresponding coxae I, II, III or IV. Normally there are three pairs of genital setae and there may be setae on the genital plate if present. The anal complex of setae is divided into either the pre-anals (pr-a) which normally consist of from one to two pairs and the post-anals (p-a) consisting of from one to five pairs of setae. When these two groups are continuous they are referred to as simply the anals (a).

Body setae may be smooth, pectinate, bipectinate, fringed, leaf-shaped, spatulate or a mere spine. Certain specialized setae within the Astigmata are sensory in function and may be referred to as omega. These are associated with the leg segments. Other specialized setae include the supracoxal, pseudo-stigmatic setae, Grandjean's organ, solenidia or rodlike setae of the tarsus. There are pores associated with the hysterosoma, which form large pigmented areas around the opisthonotal gland opening, the latter seen in many members of the genus *Tyrophagus*. Pores are well-developed in members of the genus *Histiostoma* and are said to be osmo-regulating organs.

The legs within the Astigmata may be used for walking or highly modified for clasping a host. Normally each member has eight legs, with exceptions such as the adult species, *Sphaerogastra crena* which bears only six legs. Each leg may consist of five free segments, the tarsus, tibia, genu, femur, trochanter and a coxa, the latter of which is incorporated into the ventral surface of the body. However, the five free segments are frequently reduced by fusion depending upon the particular habit of life. For example, in the genus *Labidocarpus*, legs I and II consist of modified flaplike clasping organs; in the family Acaridae the empodial claw may be connected to the pretarsus by a pair of sclerotized rods or condylophores; and caruncles may be present or absent, suckerlike, stalked or greatly reduced. Suckers may be present as in members of the genus *Tyrophagus* on the tarsus of legs IV for clasping the female. As a group, the Astigmata have lost the lateral claws and the median is the only one developed. However, the pretarsus often is well developed and carries at its apex a bell-shaped sucker. The pretarsus may be greatly elongated as in the genus *Ensliniella* or very short as in members of the genus *Sennertia*. The ventral coxal apodemes are important structures in the classification of species within the Astigmata. Systematists pay keen attention to the apodemes, which project inward, as to whether they are closed or opened. Also used extensively in classification are the tarsal solenidia which are sensory setae and are bacilliform. However, those on the tibia and genu are setiform. The tibial solenidion normally are long and whiplike on all legs.

The genital orifice is either a longitudinal slit, which may have branches and appear as an inverted "Y" or a simple transverse slit. The opening is normally located in the intercoxal area, often associated with two pairs of genital suckers. The male has a sclerotized aedeagus with shape and size utilized in determination of the species. The anal opening is ventral and, in the male of some species, is flanked by a pair of anal suckers.

There are groups which may have heteromorphic males and the deutonymph may lack mouthparts being referred to as a hypopus or wandering nymph. Hypopi are often provided with a conspicuous ventral opisthosomal field of suckers. These hypopi may be free-living or more commonly associated with either mammals or invertebrates. The hypopial stage may occur between the protonymph and tritonymph as in the family Acaridae. This hypopial or deutonymphal stage aids in distributing the species and in survival under stress conditions. There are two kinds of hypopi: those that are capable of moving freely and clinging or attaching themselves to arthropods and mammals; and the inert hypopi which are incapable of movement and must depend on the movement of air currents as a means of transportation. In many cases the hypopi are not transported from one area to another but merely await the coming of more favorable environmental conditions. Some species are capable of forming both an active or an inert hypopus. The hypopi bear no resemblance to the adults and thus make the classification of groups that have hypopi very difficult. In many families it is this hypopial stage that the entire classification is based upon.

Internal structures include the pharynx, a long narrow esophagus which is an anterior extension of the midgut region and is sharply differentiated from the pharynx. The esopha-gus opens into the stomach which is associated with several caeca in some species. Posteriorly, the stomach runs into the midgut and then to the rectum. Between the stomach and the rectum the excretory canals join the gut. The Astigmata gut is characterized by the development of the mesenteron behind the stomach into a spherical chamber, the colon, and a post colon which opens ventrally through the anal vales representing the proctodoeum. In the family Acaridae, the esophagus is narrow, long, and thin. It may have a lining of chitin and may project into the cavity of the lumen. Normally there are only two caeca. However, they are known to be absent in some families. The salivary glands lie in the podosoma.

In the male reproductive system, the testes is a paired structure. The female may possess a structure called a bursa copulatrix used as an aid in classification of species.

The nervous system is located near the esophagus which it surrounds. There is a well-developed central nervous system consisting of a brain and radiating ganglia. The lower portion of the brain, the subesophageal ganglia, tends to be responsible for the innervation of the legs, digestion, muscular, and genital systems. The mouthparts, however, appear to be regulated by the supraesophageal ganglia or dorsal region of the brain.

There is a flat ostiate heart in the dorsal region of the body in the anterior section of the gnathosomal region.

The body of a cryptostigmatid (Figs. 7, 8 & 9) mite is divided into two parts: the prodorsum and the notogaster. The prodorsum covers the propodosoma; the notogaster covers the hysterosoma. The propodosoma may be folded, may be mobile but not capable of folding, or may be firmly fused and immobile into the hysterosoma. The prodorsum contains four to six pairs of setae. These consist of a pair of sensilli (pseudostigmatal organs), interlamellar setae, lamellar setae, rostral setae, posterior exostigmatal setae, and anterior exostigmatal setae. Sensilli are situated in cuplike excrescences, the bothrydium (pseudostigma) may be present or absent altogether, the rostrum may bear teeth or indentations. The lamellae extend from the base of the prodorsum. There are two types: flat, lath-shaped, with a horizontal extension; and riblike, or horizontal. The latter are referred to as costulae. The prodorsum is separated from the notogaster. The notogaster is undivided in most cases but may be divided by transverse sutures. Neotrichial setae are usually of a different shape than the normal setae, and they appeal generally in the posterior region of the hysterosoma. Notogastral setae may be extremely short or entirely reduced to only their points of insertions which are called alveoli. The notogaster has special organs: the areae porosae, sacculi, and pori, which are highly characteristic. Slit or dotlike openings on the surface of the area porosa are called sacculi; if there are pointlike pores, they are called pori. Anterior and lateral to the notogaster are appendages resembling a wing, the pteromorphae. Typically, the pteromorphae have a downward bending chitinous, movable lamella. Also, there is a smaller horizontal wing, which is immovable, never bending vertically downward. In some species, both the vertical and horizontal pteromorphae occur, however, in most cases one or the other is present. The humeral process, found in pycnonotic oribatids is directed forward, not laterally and never movable.

Ventrally the mandibles are chelate, rarely aciculiform, or setae-shaped and serrated. The tarsi have one or three claws, rarely only two. In the tridactylous forms, the two lateral claws are much thinner than the middle claw. The sternal region has four epimeral or coxisternal plates which may be fused together in many species. The hysterosoma bears genital and anal plates which may be extremely large, occupying the entire

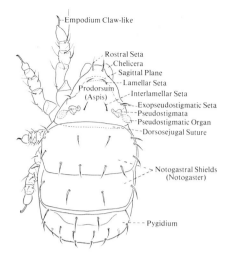

Figure 7

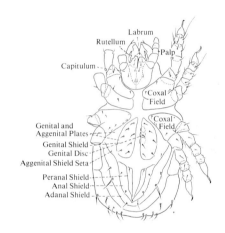

Figure 8

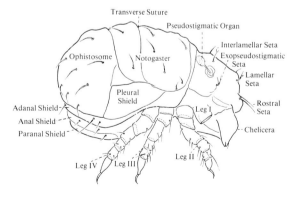

Figure 9

length of the ventral plate. The genital and anal plates may be circular, combined to form a "V," set apart, or located on separate ventral plates. The genital plates bear from four to six pairs of genital setae, the anal plates normally have two setae, and the ventral plate has four pairs of setae. The degree of coalescence of these plates and the numbers and positions of the setae are used in identification. The setae on the ventral region may at times be much reduced and are noted only by the presence of their alveoli. Important features for members of this suborder are: simple palpi without claws, each with three to five segments; well-developed sub-capitular rutellae; tarsi with one to three tarsal claws; empodium, if present, always clawlike, normally with a pair of prodorsal pseudostigmatic organs; genital and anal opening with various shaped plates, normally with genital discs.

Within the suborder Cryptostigmata, often referred to as the "beetle mites," many superfamilies have been established and placed into three supercohorts: the Palaeacari, the Oribatei inferiores and the Oribatei superiores. A worker is referred to Krantz (1978) for a list of the superfamilies and families that are contained within the three superchohorts.

Key to Major Groups
of Mites and Ticks

1a   Stigmata eight in number, four to a side, dorsolaterally on the hysterosoma, two pairs of ocelli, terminal claws present on tarsus of palp; large mites .... **Order Opilioacriformes, Suborder Notostigmata**

This order and suborder includes some of the most primitive forms of the Acari. The group as a whole would be encountered under rocks or other protected areas. They have been collected in semi-arid regions within the United States. They are large when compared with other peritreme mites. Their food habits are not well known, but they are thought to prey on other arthropods or feed on pollen grains. This group has been relegated to include only the single family Opilioacaridae following the work of Krantz 1970.

Family Opilioacaridae

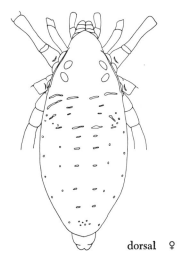

dorsal ♀

Figure 10

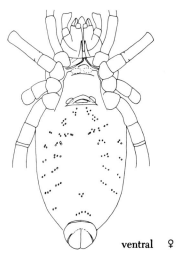

ventral ♀

Figure 11

Figures 10 and 11 *Opilioacarus texanus* (Chamberlin & Mulaik)

This species was found associated with the ground cover under rocks in the Big Bend National Park, Texas. This group of mites probably feeds on pollen grains as well as being carnivorous. They are mostly nocturnal in their feeding habits.

1b   Stigmata two in number or absent 2

2a   Stigmata located on hysterosoma: there being only one pair placed on the lateral region of the body. Order Parasitiformes 3

2b   Stigmata absent on hysterosoma: specialized propodosomal sensory organs present or absent. Order Acariformes 5

3a   Palpal tarsus with a terminal, subterminal or basal simple or fixed apotele; hypostome not modified into a piercing organ with retorse teeth 4

3b   Palpal tarsus without apotele; hypostome modified into a piercing organ with retorse teeth; tarsus I with a distinct Haller's organ (sensory pit); stigmata associated with a stigmal plate .... (p. 96) Suborder Metastigmata (the ticks)

4a   Tritosternum present, commonly with laciniae, hypostomal venter with maximum of four pairs of setae, tectum present epistome by some authors) .... (p. 30) Suborder Mesostigmata (the Peritreme mites)

This group of mites includes what are commonly referred to as the peritreme mites or mesostagmatids. The mesostagmatids comprise by far the majority of the species apt to be encountered by the beginning collector of mites in the United States. This group includes both free-living and parasitic species. The former group may be encountered in the soil, mulch, plants, food materials, nests, and as predators of other mites on animals. The parasitic species may be either external or internal parasites of mammals, birds, reptiles, insects, or other invertebrates.

The presence of a tritosternum with 1-3 laciniae on the free-living species is characteristic. However, this structure may be absent in some parasitic species. The presence of both dorsal and ventral shields along with peritremes, a tritosternum, an inner basal palpal apotele, with two, three, or four tines, and a tectum will distinguish a mesostagmatid from all other mites even though any one of the above may be missing from specific species.

4b   Tritosternum markedly reduced or absent, only rarely with laciniae, hypostomal venter with more than four pairs of setae, tectum absent .... Suborder Tetrastigmata

These mites are not as yet recorded from the United States. They are heavily sclerotized and like the Notostigmata, are large mites. The group has in the past been referred to as the Holothyroidea. The lack of a well developed tritosternum and the presence of more than four pairs of setae ventrally on the gnathosoma distinguishes this group from other mites.

They are considered to be predators and have been recorded from New Zealand, Australia, New Guinea, Ceylon, Mauritius and the Seychelles. Members of the Tetrastigmata are grouped into a single family the Holotyri-dae.

5a Beetle-like mites; usually heavily sclero-tized; rutellae present; generally with a

pair of pseudostigmatic organs, empo-dium when present claw like; tarsi 1-3 claws; palpi composed of 3-5 segments .... (p. 285) Suborder Cryptostig-mata (the Beetle-mites)

5b Rutellae usually absent; not beetle-like in shape 6

6a Normally with opisthonotal "glands"; if opisthonotal glands absent than without highly developed propodosomal sensory structures; tracheal system absent; true claws absent; emodium either claw or sucker-like, normally inserted on a dis-tinct pretarsus; genital opening normal-ly in the shape of an inverted U, V, or Y or occasionally transverse .... (p. 206) Suborder Astigmata

6b Without opisthonotal "glands"; empo-dial structure commonly pad-like, rayed, or broadly membranous on at least some of the legs; propodosomal sensory struc-tures present or absent; palpi often mod-ified into a thumb-claw process; stigmata present or absent, when present associ-ated with anterior region of body; some members may be worm-like in body shape and process a reduction in the normal compliment of legs ....(p. 110) Suborder Prostigmata

Keys to the Common
Peritreme Mites
of the Mesostigmata

1a   Female with one primary epigynial shield (genital orifice or genital cover by other authors), chelicerae without filamentous or tree-like excrescences on movable digit. Male with or without a spermatophoral process ....
.... Supercohort Monogynasida 2

1b   Female with three primary epigynial shields, these being made up of two latigynials and one mesogynial shield (paired laterals and the genital by some authors). These shields may be fused or free to open like a trapdoor. Chelicerae with filamentous excrescences, male without spermatophoral process ....
.... Supercohort Trigynaspida 35

2a   Female genital setae present 3

2b   Female genital setae absent 31

3a   Sternal shield divided or deeply incised behind sternal setae II. Epigynial shield containing one, three, or several pairs of genital setae. Dorsal shield either entire or divided into several shields, marginal plate-like shields present. Male genital opening between coxae II, III, or IV.

Chelicerae without spermatodactyl ......
Cohort Sejina, Superfamily Sejoidea 4

3b   Sternal shield entire. One pair of genital setae associated with epigynial shield, either located on the epigynial shield or close to it (when genital setae are not on epigynial shield they may be distinguished from other setae by the presence of a pair of pores associated with them, when epigynial shield is fused with ventral shield to form a genitiventral shield the genital setae are those located in the podosomal region). Male genital opening at anterior margin of sternal shield or within it; chelicerae with or without spermatodactyl 6

4a   Sternal shield deeply incised (fragmental); with many pairs of genital setae; both male and female with 2 pair of large medium dorsal shields and 2 pair of smaller lateral shields ....
.... (Figs. 12-13) Uropodellidae

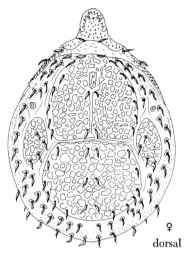

Figure 12

Figure 14

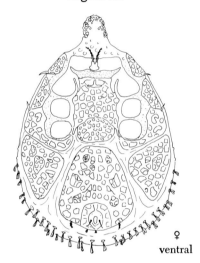

Figure 13

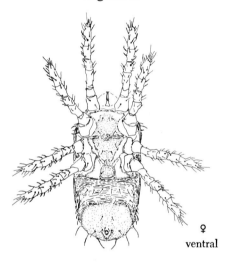

Figure 15

Figures 14 and 15 *Microgynium incisum* Krantz

The general shape of this species is rectangular with numerous pectinate vertex setae. There are three median dorsal shields, the anterior dorsal shield which is the largest and two smaller shields posteriorly on the hysterosoma.

This species was collected from hemlock litter, Odell Lake, 28 miles east of Oakridge, Oregon. This is the only recorded species from the United States. Two other species have been described, *M. rectangulatum* and *M. trunicola* from Scandinavia.

These mites are forest dwellers and are known to feed on fungi.

4b Sternal shield divided 5

5a Sternal shield divided into two subequal portions, both the sternal and epigynial weakly sclerotized, three dorsal shields, anterior largest, posterior smaller .... **Microgyniidae**

5b With from one to seven dorsal shields, normally dorsum with network of narrow ridges .... Sejidae

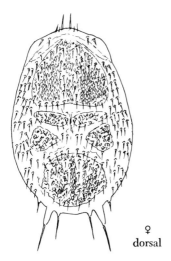

♀
dorsal

Figure 16

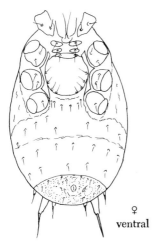

♀
ventral

Figure 17

Figures 16 and 17 *Sejus* sp.

Several species of the genus *Sejus* have been collected in the United States. However, members of this family are rather uncommon and are found associated with litter and fallen logs of forest and orchard soils. Members have

a single epigynial shield with from one to several pairs of epigynial setae in combination with from one to several dorsal shields and a network of narrow ridges or platelets.

6a Sternal shield of female fused with endopodal, parapodal, and peritremal shields and partially encircling epigynial shield .... Cohort Uropodina 34

6b Sternal shield of female not fusing with other accessory shields of venter to form a plate that encircles epigynial shield. Male genital aperature in front of sternal plate or within it, chelicerae often modified to transfer spermatopheres .... Cohort Gamasina, Superfamily Parasitoidea 7

This group of peritreme mites is an important group and it contains most of the common mesostigmatid mites collected. The name of the super-family should not mislead a worker in assuming that only parasitic species are to be encountered within this superfamily due to the prefix parasito—as it contains both free-living and parasitic forms.

7a Female metasternal shields forming an inverted "V" into which the triangular epigynial shield fits. Palpal apotele three-tined. Male with apophyses on leg II. Free-living .... Parasitidae

Representatives of this family are commonly associated with litter and humus. The adults are heavily sclerotized, fast moving and are distinguished by the structure of the female genital region and chelicerae of the male. The immature stages are commonly found on beetles and other insects, which they use as a means of transportation. The members of this family have not been worked on in the United States and are difficult to classify. Both the female and male are needed for specific identification.

7b   **Female metasternal shields not forming an inverted "V" into which epigynial shield fits, palpal apotele not always three-tined** 8

8a   **Palpal apotele associated with an inflated lamelliform hyaline flap. Hypostome produced into wide expansive hyaline structure in females (mustache-like fringe by some authors). Free-living** .... (p. 54) **Veigaiidae**

8b   **Palpal apotele not associated with an inflated lamelliform hyaline flap. Hypostome with or without expansive hyaline structure** 9

9a   **Peritreme associated with stigmata looped around the opening of the stigmata joining the stigmata posteriorly. Legs I and II as a rule without apoteles. Epigynial shield with a pair of well developed internal sclerotic bars. Free-living** .... (Fig. 18) **Macrochelidae**

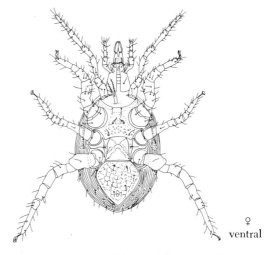

♀ ventral

Figure 18

The mites of this family may be distinguished from related families by the strongly looped peritreme at the site of the stigma in combination with the absence of claws on the first pair of legs. The macrochelids are widely distributed in a variety of habitats. They are mostly free-living and one species *Macrocheles musae-domesticae* (Scopoli) has been found to feed on the eggs of the housefly. Because of this these mites are commonly found in large numbers associated with cow manure.

9b   **Peritreme not looped around stigmata, joining the stigmata anteriorly or laterally** **10**

10a   **Brush-like setae present on movable digit of chelicera. Palpal apotele three-tined (central tine often spatulate) Free-living** .... (p. 56) **Parholaspidae**

10b   **Brush-like setae absent on movable digit of chelicera. Palpal apotele two or three-tined** **11**

11a Sternal shield extending to coxae IV, longer than wide; parapodal, peritremal and metopodal shields fused and extending beyond posterior portion of coxae IV, free-living or associated with insects .... **Pachylaelapidae**

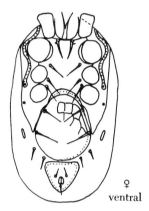

♀
ventral

Figure 19

Figure 19 *Pachylaelaps* sp.

The pachylaelaptids are distinguished from other parasitoids by the sclerotization of the venter. The inter-coxal region in the female is covered by the fused parapodal and metopodal shields. This is compounded by the fusion of this region by the expodal-peritrematal shield. Leg II of the male is heavily armed with spines, tarsus II in the female may also be provided with stout spurs.

Members of the genus *Pachylaelaps* are found associated with members of the insect order Coleoptera (beetles), especially the family Scarabaeidae. However, other species of this genus are found associated with soil mulch, or humus. Members of the genus *Sphaerolaelaps* are myrmecophilous (ant-loving). The relationship of members of this family associated with insects is non-parasitic and is referred to as phorsey (phoretic). Members of the genus *Pachylaelaps* have been collected from the state of Oregon.

11b Sternal shield rarely extending to middle of coxae IV, if reaching coxae IV then associated with discs or suckers on the opisthosoma **12**

12a Opisthosomal venter with two well-developed disc-like suckers; sternal shield reduced. Associated with snakes and invertebrates .... **Heterozerconidae**

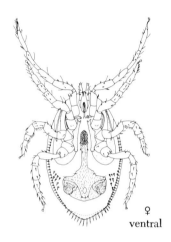

♀
ventral

Figure 20

Figure 20 *Heterozercon oudemansi* Finnegan

This family contains members that are sometimes found associated with reptiles, birds, mammals and insects. In the case of insects it is believed to be a phoretic relationship. The genus *Heterozercon* contains a number of species, *H. oudemansi* Finnegan was reported from the snake *Epicrates cenchria* from South America. An undescribed species has been reported from Florida by Krantz (1970). The genus *Heterozercon* is based on the presence of the sucker-like disk located on each side of the anus.

Members of the family are probably tropical or subtropical in distribution and it could be anticipated that additional collecting in the southern United States will establish the presence of this family in the United States.

12b **Opisthosomal venter without two well-developed disc-like suckers** **13**

13a **Apotele of palpal tarsus with three tines; corniculi undivided distally; chelae, chelate-dentate or, if fixed digit reduced, with movable digit dentate** **14**

13b **Apotele of palpal tarsus usually with two tines (if three-tined, then corniculi divided distally)** **16**

14a **Dorsal shield divided, claws present or absent, tarsal setae three-pronged. Associated with soil, litter, moss, dung, and the interstitial regions of beaches ....**
.... **Rhodacaridae**

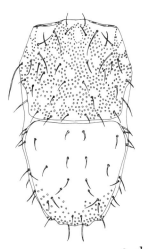

♀ dorsal

Figure 21

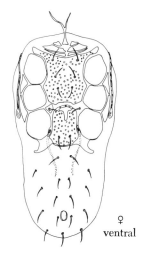

♀ ventral

Figure 22

Figures 21 and 22 *Psammonsella nobskae*

This family is often referred to as a heterogenous assemblage of free-living mites. They may be associated with insects or other invertebrates (phoresy), found in the soil, humus, moss, decaying vegetation, dung, compost heaps, nests of and on small mammals, pond algae, etc.

The rhodacarids are poorly known. There is a wealth of species and the beginner as well as most professionals would have a difficult time placing individuals into species catagories.

Members of the genus *Rhodacarus* Oudemans have been collected from Oregon and Massachusetts, those from the latter state were in association with the interstitial area of Nobaska Beach southeast of Woods Hole. Members of *Rhodacarus* have been collected from litter at Cottonwood, South Dakota. Species which appear to belong to the genus *Saprolaelaps* Leithner have been collected from dung at Brookings, South Dakota. It is interesting to note that *Saprolaelaps subtilis* was recorded from Nobaska Beach, Massachusetts.

The following species of interstitial species of Rhodacaridae have been recorded from the United States (Haq 1965): *Rhodacarus pallidas, Rhodacaropsis inexpectatus, Saprolaelaps subtilis, Cyrtolaelaps (Gamaselliphis)* sp. and *Psammonsella nobskae.*

14b Dorsal shield entire (if abbreviated, with mesonotal scutellae present) 15

15a Legs one antenniform, lacking pretarsus, terminating in either one or two long whip-like setae; palpal claw with three tines; dorsal shield tuberculated .... **Podocinidae**

Members of this family are free-living with very long legs for running. The tarsus of leg I has two long whiplike setae; the dorsum has two pores on the posterior region of the body and is covered with a network of ornamentation.

15b Legs one terminating in a series of short simple hairs; dorsal plate entire, palpal claw three-tined, chelicerae dentate .... **Arctacaridae**

The members of this family are free-living forms with legs I widely separated from legs II. The epignial shield is without setae and with mesonotal scutellae. Little is known regarding their life history.

16a Epigynial shield truncate at posterior portion, may be fused with a ventrianal shield, if rounded then anal shield will not be triangular in shape 17

16b Epigynial shield rounded or pointed, if rounded anal shield is triangular in shape. Parasites of various animal groups or free-living 22

17a Sternal setae consisting of two pairs on sternal shield; corniculi forked distally; anterior rostral setae thickened; dorsal shield entire in both sexes and bears 24 to 29 pairs of simple or plumose setae .... **Ameroseiidae**

Members of the mite family Ameroseiidae have been collected from flowers where they are believed to feed on pollen There is a phoretic relationship between species of this family and certain bees.

17b Sternal setae consisting of three or four pairs on sternal shield, without combination of above characteristics 18

18a Metasternal setae on separate platelets or inserted free on integument adjacent to sternal shield. Sternal shield with two to three pairs of setae; dorsal shield entire, incised, or divided medially 19

18b Metasternal setae inserted on the posterolateral region of sternal shield, with four pairs of setae; dorsal shield divided medially; free-living or associated with insects .... **(p. 80) Digamasellidae**

19a Dorsal setae numbering less than twenty pairs **20**

19b Dorsal setae numbering more than twenty pairs **21**

20a Fixed chelae developed; ventral and anal plates fused, anal opening subterminal. Predators, associated with shrubs, trees, grass, in some species feeding on fungus or plant tissues .... **(p. 69) Phytoseiidae**

20b Fixed chelae vestigial; anal opening terminal; with only three pairs of sternal setae, two pairs of which may or may not be on sternal plate, three pairs always on integument. Parasites of insects (moths) .... **Otopheidomenidae**

Species of this family are found as parasites of phalaenid moths. They infest the tympanic cavities and can cause deafness in its host.

21a Peritremes normally developed not extending far beyond coxae IV; with a ventrianal or an anal shield, latter rarely triangular. Free-living or phoetically associated with insects or birds ....
.... **(p. 74) Ascidae**

21b Peritremes greatly reduced in length not extending beyond coxae IV posteriorly and coxae III anteriorly; with a transverse row of four large crescent-shaped pores on posterior edge of opisthonotal shield. Free-living .... **(p. 82) Zerconidae**

22a Chelicerae large, hooked, set in an obvious camerostome. Parasites of bats ....
.... **Spelaeorhynchidae**

This family of mites was at one time placed in the Metastigmata (ticks). In a recent study of this group it has been established that though members of this family do not agree perfectly with either Metastigmata or Mesostigmata they nevertheless are much more closely related to the latter group. Members of this family are not at present recorded from the United States.

22b Without camerostome or hooked chelicerae **23**

23a Peritremes absent or vestigial; respiratory tract parasites of mammals, birds and snakes **24**

23b Peritremes present, they may be small but completely formed. External parasites of vertebrates, invertebrates or free-living **26**

24a Tritosternum normally present; sternal and epigynial plates well developed; sternal setae small or absent. Stigmata dorso- or latero- ventral. Respiratory tract parasites of snakes .... **Entonyssidae**

Members of this family are parasites of snakes. They are long-legged mites with an undivided dorsal plate with the tritosternum lacking. The peritreme is very short or entirely lacking in some species. They are found internally and are hard to see as they are poorly sclerotized.

24b Tritosternum normally absent; sternal and epigynial plates normally vestigial, one or both frequently absent **25**

25a Epigynial shield lacking or vestigial; sternal shield sometimes lacking; spiracles ventral. Respiratory tract or lung parasites of mammals ....
.... **(p. 44) Halarachnidae**

25b Epigynial shield present but reduced. Stigmata dorso- or lateroventral; sternal shield reduced or lacking, sternal setae present and distinct in podosomal region. Respiratory tract parasites of birds
.... **(p. 46) Rhinonyssidae**

26a Tritosternum absent or vestigial; legs I-IV similar in thickness. Stigmata usually dorsal, peritremes long. Plates of ventral surface often reduced. Parasites of bats .... (p. 92) Spinturnicidae

26b Tritosternum present, well developed with laciniae 27

27a Chelicerae not well developed, stylettiform in structure, endentate, sometimes with reduced, vestigial transparent teeth. Corniculi usually indistinct 28

27b Chelicerae well developed, strongly dentate, corniculi well defined 29

28a Chelicerae attenuated, stylettiform, distal cheliceral segment elongate, exceeding length of basal segment; movable digit minute. Palpal trochanter usually with spur. Parasites of rodents and birds .... (p. 60) Dermanyssidae

28b Chelicerae not attenuated, stylettiform; chelae normal, well developed, edentate. Parasites of mammals, birds and reptiles .... (p. 63) Macronyssidae

29a Corniculi elongate, apically beset with barbed (harpoon-like) chelicerae with large, recurved teeth. Coxae sometimes with blunt spur-like setae. Ectoparasites of snakes .... Ixodorhynchidae

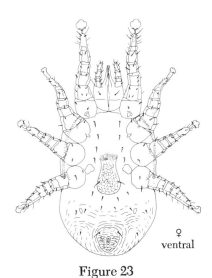

♀
ventral

Figure 23

Figure 23 *Ixodorhynchus liponyssoides* Ewing

Dorsal shield divided by a transverse suture into two subequal parts; peritreme long, extending beyond anterior coxae; sternal plate large, anal plate oval; legs short; chelicerae with three or more recurved teeth or single chela.

29b Corniculi not elongated, not apically beset with barbed chelicerae 30

30a Tectum long, tongue-like, this may include the central portion of the epistome. Epigynial shield parallel-sided. Peritremal shields extend beyond coxae IV. Commonly found associated with insects or crustaceans .... Eviphididae

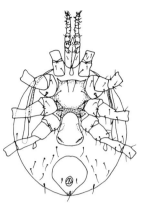

Figure 24

Figure 24 *Eviphis* sp.

The species included in this family are known to be predaceous and are many times found to be associated with other invertebrates. The males are distinguished by the presence of a sterniti-genital shield separate from an anal shield. The females are similar to members of the family Ascidae.

30b   **Tectum normal, not long and tongue-like. Epigynial plate distinct, expanded, drop-shaped. Peritremal shield normally not extending beyond coaxe IV. Genu and tibia I each with two anterolateral setae. Free-living, parasitic on birds or mammals or associated with inverte-brates** .... **(p. 84) Laelapidae**

31a   **Sternal shield without lateral, intercoxal projections. Adanal setae more than one-half the length of the body** ....
.... **Diarthrophillidae**

This group of mites is associated with the common passalid beetle. The species *Diarth-rophallus quercus* Pearse and Wharton (fig. 25) was collected from a beetle at Durham, North Carolina. This is the only recorded species from North America.

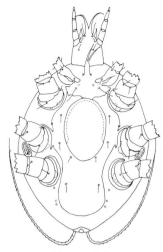

Figure 25

31b   **Sternal shield fused with endopodals and projecting laterally between coxae. Adanal setae normal** **32**

32a   **Coxae I usually contiguous and covering base of tritosternum. Foveolae pedales present or absent. If coxae I are widely separated, then not covering tritosternal base. The foveolae pedales or peritreme-bearing humeral projection would be present if above is case** ....
.... **Superfamily Uropodoidea 33**

This superfamily contains the following families that fit the above description: Thinozerconidae, Dinychidae, Eutrachytidae, Coxequesomidae - Planodiscidae - Circocyllibanidae complex, Metagynellidae and Uropodidae.

    The family Uropodidae is commonly found in earthworm beds in the United States. Members of the family Thinozerconidae are free-living forms with only a few species having been described. Dinychidae may contain species that are predatory. Eutrachytidae and Metagynellidae are free-living, in tree holes (*Metagynella*). The complex families Coxequesomidae-Planodiscidae-Circocyllibanidae are associated with the dreaded army ants of the Neotropical realm. They are of great interest in their peculiar relationship with these ants.

32b Coxae I widely separated; tritosternal base two times as long as wide. Tectum short, broadly triangular with large irregular teeth. Dorsal marginal shields absent. Epistome short, broadly triangular .... **Protodinychidae**

This family contains a single genus with species described from flood water debris in England. However, an undescribed species is known to occur in North America, associated with a beetle which was found with beaver lodges.

33a Coxae I not widely separated; tritosternal base normal, tectum normal **34**

33b Palpi 4-segmented (tibia and tarsus fused) corniculi reaching to or beyond level of distal margins of palpal femur, more than twice as long as broad. Perigenital rim present. Legs I without claws .... **Polyaspidae**

Species of this family are found associated with tree-hole microhabitats and are thought to have a phoretic stage for distribution. Species have been recorded from Florida northward to Missouri and Illinois in the United States and north to Canada. When encountered they are usually found in large numbers.

34a Palpi 5-segmented; corniculi less than twice as long as broad. Perigenital rim absent. Legs with or without claws .... **Trachytidae**

The members of the family Trachytidae are heavily armored forms similar to the family Polyaspidae but differ in that leg I contains small claws and the epigynial shield of the female distinctly articulated and without a perigenital rim.

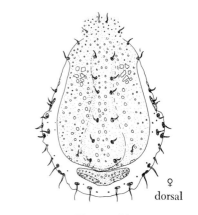

♀
dorsal

Figure 26

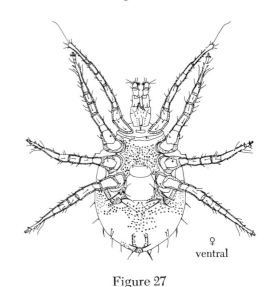

♀
ventral

Figure 27

Figures 26 and 27 *Caminella peraphora*

34b Tectum long, narrow, hyaline and spinose. Dorsal marginal setae on platelets or elongate marginal shields. Epigynial shield located between coxae II-IV, lacking setae in some genera. Sternal pores III in region of coxae. Metapodal plates fused with ventral plate or separated by a fine metapodal line .... Uropodidae

35a Genital lateral shields elongated. These shields overlap the reduced mesogynial shield. Two sub-equal dorsal shields present. These may appear to be fused together but the line of fusion is always visible. Tritosternal laciniae fused as one for more than half its length, at times only the apex separated or at times not separated at all. Epistome with anterior projections or serrations, lacking a keel. Normally found in association with arthropods ....
.... Cercomegistidae

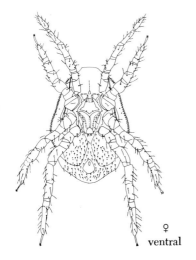

Figure 28

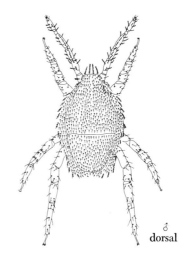

Figure 29

Figures 28 and 29 *Cercoleipus coelonotus* Kinn

Shape oblong, oval; tritosternum well-developed; latigynial shield elongated, curved, overlapping mesogynial plate posteriorly; male with cerci, small, heavily sclerotized; elevated medial postero-dorsal plate bears three pairs of stout setae. Collected from the abandoned mines of *Ips confusus,* bark beetles in *Pinus monophylla,* from California.

35b Genital lateral shield not overlapping reduced mesogynial shield. Tritosternum normal, epistome simple, pretarsal elements absent 36

36a Sternogynial shield or shields present and distinct, bearing sternal pores III; ventral and metapodal shields separated by at least two narrow bands of soft integument; anal shield separate from or fused with ventral shield. First sternal setae on sternal shield, or in soft integument, but not in a tetartosternum, without jugular shields ....
.... Megisthanidae

Large mites, found in association with the patent leather beetle, *Polilius disjunctus,* in the southeastern United States. Other species have been found in other parts of the world in association with passalid, dung beetles and other ground beetles. A *Megisthaus* sp. is known to exist in South Dakota associated with dung beetles.

36b Sternogynial shield not clearly developed; sternal pore III on sternal or metasternal shield or not clearly marked 37

37a Sternal setae I always located on the sternal shield; vaginal sclerites well developed, usually with heads. Chelicerae well developed with dendritic or brushlike excrescences, with large proximal tooth on movable digit. Associated with insects, normally passalid beetles 38

37b Sternal setae I free not located on sternal shield, placed on soft integument of body or on weakly sclerotized sternal shield. Vaginal sclerites reduced, without heads, and with a bow-shaped base. Chelicerae tapered, with filamentous excrescences, often edentate or with numerous minute teeth 41

38a Latigynial shields well developed, free medially, hinged or fused with ventral shield posteriorly; mesogynial shield well developed or absent 39

38b Latigynial shields partially or entirely fused medially; mesogynial shield never well developed 40

39a Latigynial shields well developed often extending posteriorly beyond coxae;

fused, not hinged to ventral shield. Associated with snakes and beetles .... **Schizogynidae**

Members of this family are associated with bark beetles in the western United States.

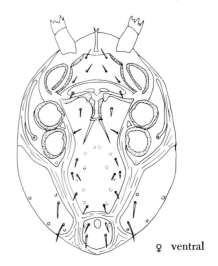

♀ ventral

Figure 30

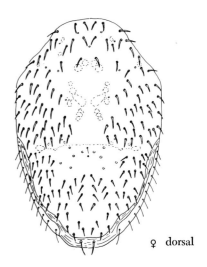

♀ dorsal

Figure 31

Figures 30 and 31 *Choriarchus reginus* Kinn

Metasternal shields not fused with sternal shield, but fused with one another to form a single plate; metapodal plates free.

Collected from galleries of *Pseudohylesinus grandis* infesting *Pseudotsuga menzieii*, *Phloeosinus punctatus* infesting *Libocedrus*, *P. sequoiae sempervirens*, all from California.

39b   Latigynial shields well developed, elongate not extending posteriorly beyond hind margins of coxae III and never to middle of coxae IV, hinged to ventral shield, mesogynial shield usually hinged, sometimes reduced and fused with ventral shield or absent .... **Dipogyniidae**

40a   Latigynial and mesogynial shields fused with ventral shield and with each other except along anterior margins .... **Euzerconidae**

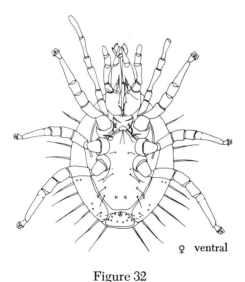

♀ ventral

Figure 32

Figure 32   *Euzercon latus* (Banks)

This species is found associated with the passalid beetle, *Popilius disjunctus*, in North Carolina and is to be found throughout the range of its host.

40b   Latigynial shields entirely fused with ventral shield and with each other, except for a shallow indentation anteriorly; no remnant of shield present .... **Celaenopsidae**

Members of this family are known to occur in association with bark beetles in Colorado.

41a   Latigynial shield sclerotized along anterior edges; mesogynial shield reduced, unhinged, and almost entirely coalesced with the latigynials. Chelicerae with minute teeth. Associated with insects .... **Antennophoridae**

These mites are found in association with ants. Three species have been recorded from the United States.

41b   Latigynial shields sometimes with a sclerotized seta-bearing patch; mesogynial shield hypertrophied, hinged and often bearing setae. Chelicerae with a few distal teeth. Associated with insects and myriapods .... **Parantonnulidae**

Members of this family have been recorded from northcentral United States. Undescribed species have been found associated with carrion. Other reports associate them with carabid beetles.

Family Halarachnidae

This family contains mites that are parasitic in the respiratory tracts of mammals. Their body shape is worm-like. This family is very close in structure to other internal parasitic groups such as the Rhinonyssidae, and Entonyssidae. In the work of Evans and Till (1966) these endoparasitic forms are classified as subfamilies of the family Dermanyssidae. The author will not attempt to justify the treatment of these groups at the family level as in all probability because of the similarity in general morphological structures they should indeed be treated as sub-families.

1a   Parasitic in the respiratory system of of marine vertebrates (Pinnipedia—hair seals) 2

1b   Parasitic in respiratory system of terrestrial vertebrates 3

2a   Male and female with more than four pairs of setae dorsally between the posterior margin of the dorsal plate and anal plate and also ventrally between the level of the fourth pair of coxae and the anal plate .... Genus *Halarachne*

2b   Male and female with only two pairs of setae in the above areas, abdomen of the gravid female greatly elongated in most species, parasites of fur seals and walruses .... Genus *Orthohalarachne*

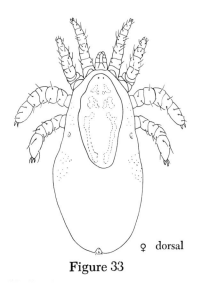

♀ dorsal

Figure 33

Figure 33   *Halarachne sp.*

Female rather stout bodied, about half as wide as long, dorsum without setae, sternal plate with four setae, and three on the anal plate, all very small, legs slender, a very definite chitinized band connects spiracle with genital slit.

Collected from *Mirounga angustirostris* on the Pacific Coast, from Baja, California to Washington State.

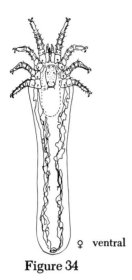

♀ ventral

Figure 34

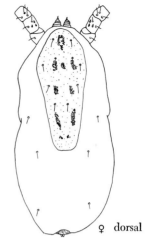

♀ dorsal

Figure 35

Figure 34 *Orthohalarachne attenuata* Banks

Dorsal shield, peritreme extending to coxa III, sternal plate small, narrow, emarginate in front and a pointed tip, legs I-IV are sub-equal, legs I and II stouter than III and IV.

Collected from *Callorhinus ursinus* and *Eumetopias jubata* from coast of Oregon.

3a   Palp with only four movable segments, the palpal tibia and palpal tarsus being fused, combined length of the movable segments of the palp much less than the fused coxae .... Genus *Pneumonyssus*

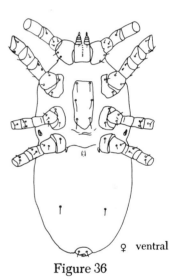

♀ ventral

Figure 36

Figures 35 and 36 *Pneumonyssus simicola* Banks

Palpi with three or four segments, claws of tarsus I of both sexes developed about as much as those of tarsus III, palpus with terminal pair of long setae, without long lateral basal setae on palpal tarsus, palpus less than half as long as the width of the basis capituli, claws of tarsus I measure approximately 20 microns in straight line from base to apex.

Collected from the Rhesus monkey commonly used for medical research throughout the United States.

3b **Palp with five movable segments, modified forked palpal claw lacking, peritremes slender and nearly as long as the diameter of coxa IV, no constriction between podosoma and opisthosoma, parasitic in dogs and wart hogs. ....**
.... Genus _Pneumonyssoides_

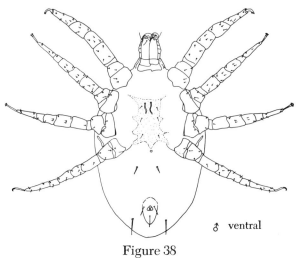

δ ventral

Figure 38

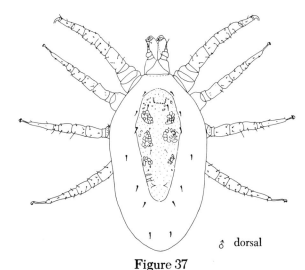

δ dorsal

Figure 37

Figures 37 and 38 _Pneumonyssoides caninum_ (Chandler and Rube)

Palpi with five segments, pretarsi of legs II, III and IV pedicellate, tarsal claws of leg I distinctly larger than those of leg III, last palpal segment an elongate tibio-tarsus, first tarsal claws approximately half as long as tibia I, peritreme over three times as long as broad, metasternal setae absent.

Collected from dogs in Michigan and California. This parasite could be expected to be found wherever dogs are found.

Family Rhinonyssidae

The mites of the family Rhinonyssidae are internal parasites of the respiratory tract of birds. Nasal mites have been recovered from most of the orders of birds. They are weakly sclerotized with a reduced dorsal shield, the tritosternum is absent and they have short peritremes. This family is classified as a subfamily of Dermanyssidae by Evans and Till (1966) in their study of the British Dermanyssidae.

1a **Chelate portion of the chelicerae forming one-tenth or less of their total length**
.. 2

1b **Chelate portion of the chelicerae forming one-seventh or more of their total length 6**

2a Stigma without peritreme 3
Stigma with short peritreme that is elongate or circular. 4

3a Mouthparts ventrally directed, not entirely visible from above, when both fore coxae are directed forward, they meet above the gnathosoma, stigma without peritreme .... **Genus *Sternostoma***

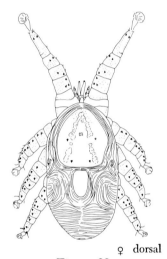

♀ dorsal

Figure 39

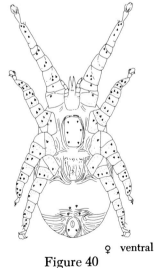

♀ ventral

Figure 40

Figures 39 and 40 *Sternostoma boydi* Strandtmann

Small whitish in color, elliptical, podosomal shield large, one pair of small setae near posterior margin, opisthosomal shield less than half as long as podosomal, devoid of setae, peritreme lacking, no dorsal setae, genitoventral plate long as wide, no genital setae, legs I not over 1.3 times as long as leg IV, tarsi II-IV with elongate setae, adanal alveoli posterior to anus.

Collected from ring-billed gull, laughing gull, rudy turnstone, and the sanderling in Texas.

3b Mouthparts apical, opisthosomal setae very thick, heavy, blunt .... **Genus *Cas***

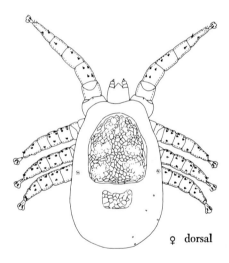

♀ dorsal

Figure 41

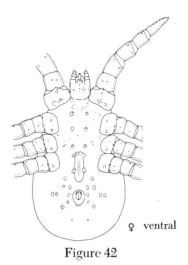

♀ ventral

Figure 42

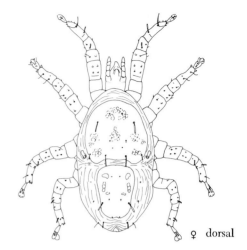

♀ dorsal

Figure 43

Figures 41 and 42 *Cas angrensis* (Castro)

Without peritreme, chelae minute, mouthparts entirely visible from above, legs I longer than II, III, IV, with two dorsal plates.

Collected from *Iridoprocne bicolor, Petrochelidon fulva, Petrochelidon pyrrhonota* and *Progne sulis,* all swallows in Texas.

4a   Chelicerae bulbous basally **5**

4b   Chelicerae of uniform diameter throughout, not bulbous at base with two dorsal plates, one podosomal and one opisthosomal, chelicerae of uniform diameter throughout ....
.... **Genus *Paraneonyssus***

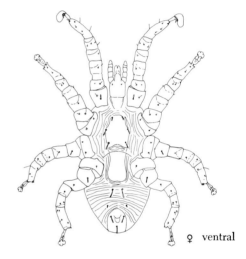

♀ ventral

Figure 44

Figures 43 and 44   *Paraneonyssus capitatus* Strandtmann

With apically expanded setae, four spur-like setae on the apices of tarsi II-IV, sternal plate lightly developed but always distinct, epigynial plate broad and bears a pair of minute setules, large anterior dorsal plate, opisthosomal plate.

5a Dorsum with two plates, one podosomal and one pygidial, latter bearing two setae on posterior margin, body elongate, slightly constricted medially .... **Genus *Ptilonyssus***

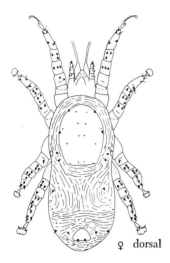

♀ dorsal

Figure 45

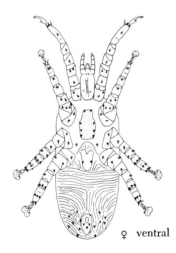

♀ ventral

Figure 46

Figures 45 and 46 *Ptilonyssus nudus* Berlese and Trouessart

Sternal plate present, weakly sclerotized, genital plate at least 1/3 as wide as it is long and rounded posteriorly, trochanters I, III and IV with two setae, podosomal plate large, weakly sclerotized, with nine pairs of minute setae, semi-circular pygidial plate, opisthosoma with 12 pairs of setae, anal plate pear-shaped with three blunt setae all posterior to anal opening, six pairs of setae in area around anal plate.

Collected in *Passer domesticus* from Texas.

5b With three dorsal plates, chelicerae heavy, short, chela large, sternal plate and tined palp setae absent, with four aligned setae on geni III .... **Genus *Tryanninyssus***

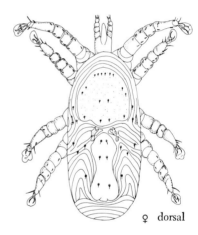

♀ dorsal

Figure 47

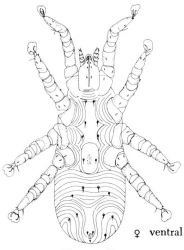

Figure 48

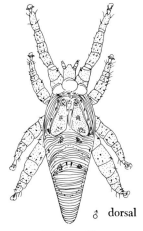

♂ dorsal

Figure 49

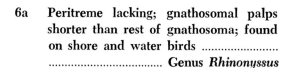

♀ ventral

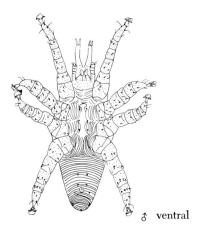

♂ ventral

Figure 50

Figures 47 and 48 *Tyranninyssus tyrannisoides* Strandtmann

Female and male with five opisthosomal plates, four small dorsal plates near anterior end of the large opisthosomal plate, this plate uniformly sclerotized for its entire length, three pairs of sternal setae, and two pairs of pores, genital plate of female ovate, with a pair of setae near posterior end flanked laterally by a pore.

Collected from Say's phoebe and the least flycatcher in Texas.

6a   Peritreme lacking; gnathosomal palps shorter than rest of gnathosoma; found on shore and water birds .... **Genus *Rhinonyssus***

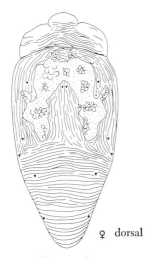

♀ dorsal

Figure 51

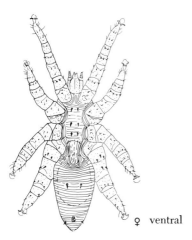

♀ ventral

Figure 52

Figures 49, 50, 51, and 52 *Rhinonyssus coniventris* Trouessart

Sternal setae short and heavy, at least the third pair appendiculate, eight ventral setae, six of which are anterior to anal pore, anal pore ventral, no trace of a sternal plate, ventral setae eight or less, no trace of an anal plate.

Collected from the ruddy turnstone, ringed plover, snowy plover, willet, lesser yellowlegs, red-backed sandpiper, sanderling,

Aleirtian sandpiper, and purple sandpiper in Texas, Florida and Alaska. Since these plovers and sandpipers migrate across the United States, this mite could be expected to be found wherever the host is found.

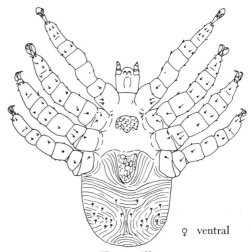

♀ ventral

Figure 53

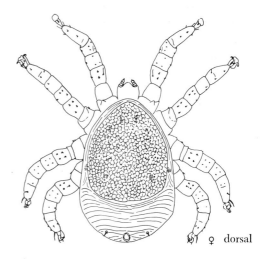

♀ dorsal

Figure 54

Figures 53 and 54 *Rhinonyssus himantoptus* Strandtmann

Anal pore dorsal, vestiges of a sternal plate, sixteen or more ventral setae, no trace of an anal plate, a large and prominent dorsal plate, loss of sternal setae.

Collected from the black-necked stilt in Texas and Florida.

6b Peritreme present, palps as long as or longer than remainder of gnathosoma
.... 7

7a Chelicera lacking the fixed digit; the movable arm well-developed; without a tritosternum, with only three pairs of sternal setae (no metasternal setae) ....
.... **Genus *Rhinoecius***

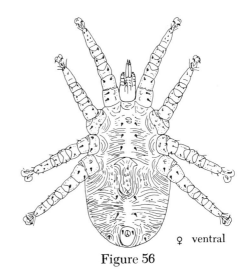

♀ ventral

Figure 56

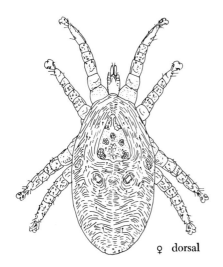

♀ dorsal

Figure 55

Figures 55 and 56   *Rhinoecius cooremani* Strandtmann

An elongated mite with a triangular podosomal shield and only two small platelets, sternal plate lacking, genital plate distinct, three pairs of sternal setae and three pairs of opisthosomal setae, male with small sternal plate, chelae composed of a short, immovable arm and a longer movable arm to which is attached a spermatophore carrier.

Collected from the barred owl, *Strix varia*, in Texas.

7b Chelicera with fixed digit present 8

8a Stigma and peritreme near the posterior part of the body, anal pore surrounded by a free membrane ....
.... **Genus *Rallinyssus***

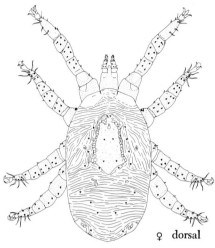

♀ dorsal

Figure 57

Figure 57 *Rallinyssus caudistigumus* Strandt-mann

8b Stigma and peritreme at normal position above or anterior of coxa IV 9

9a Dorsum with a group of small shields; mouth parts subventral, chela about one-sixth as long as the chelicera, found only in gulls and terns ...... Genus *Larinyssus*

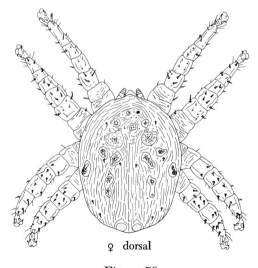

♀ dorsal

Figure 58

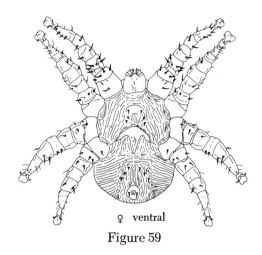

♀ ventral

Figure 59

Figures 58 and 59 *Larinyssus orbicularis* Strandtmann

9b Dorsum with two shields, chela about one-third as long as the chelicera, stigma in the normal position, chela larger, more than one-seventh the length of the chelicera .... Genus *Neonyssus*

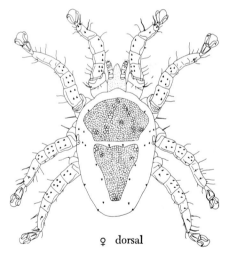

♀ dorsal

Figure 60

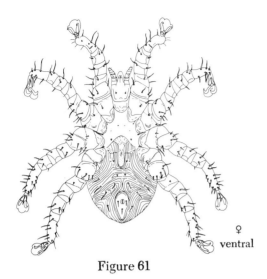

Figure 61

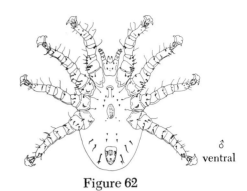

Figure 62

Figures 60, 61, and 62 *Neonyssus triangulus* Strandtmann

Posterior dorsal plate triangular, three to five pairs of long ventral setae, coxae II to IV with semicircular elevations, coxal setae unequal, three anal setae, the paired setae near the anterior margin of the pore.

Collected from the white-winged dove in Texas.

Family Veigaiidae

The members of this family are principally associated with humus, soil and moss. They are mostly predaceous on microorganisms found in the soil. Very little is known regarding the biology of members of this family. They are considered to be ovoviviparous. The males are rare and several species are known only from the description of females. According to the work of Farrier (1957) a total of 14 species have been collected in the United States, representing two genera.

1a   Ambulacri III-IV with hyaline flap between the claws not divided medially ....
.... **Genus *Gorirossia***

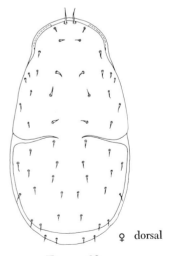

Figure 63

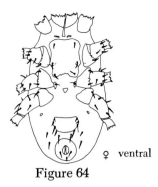

♀ ventral

Figure 64

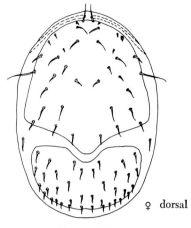

♀ dorsal

Figure 66

Figures 63 and 64 *Gorirossia whartoni* Farrier

Dorsal scutae joined medially, tectum without medial projection; fixed digit of chelicerae with two minute subapical teeth and four moderate teeth equidistant from each other; movable digit with four teeth set equidistant from each other. The epigynial plate absent; ventral plate reduced.

Collected in North Carolina from leaf mold.

1b **Ambulacri II-IV with hyaline flap between the claws divided medially, these wide and broadly rounded apically ......**
.. Genus *Veigaia*

Figures 65 and 66 *Veigaia nemorensis* (Kock)

Fixed chela with but two normal teeth, palpal genu with anteromedial seta combed; peritremal shield not joined to ventral shield.

Collected in New Hampshire from forest litter.

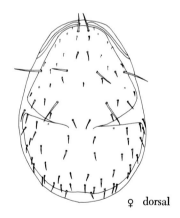

♀ dorsal

Figure 67

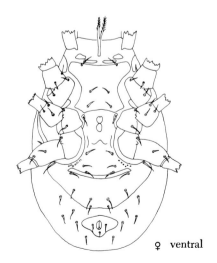

♀ ventral

Figure 65

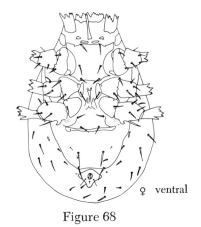

Figure 68

Figure 67 and 68 *Veigaia uncata* Farrier

Femur IV with a distinct distally directed hook, palpal genu with posteromedial seta without teeth; palpal genu with antermedial seta without teeth, simple, dorsal scutae only partially divided by a transverse suture leaving scuta joined medially.

Collected in North Carolina from leaf mold.

Family Parholaspidae

The members of this family have in the past been a part of the family Macrochelidae and listed as a subfamily. However, recent work tended to support the advancement of this group of mites to a familial concept. These mites are entirely free-living in habit and are associated with the microflora of the soil. They are distinguished from the Macrochelidae in not having the peritreme looped proximally, joining the stigma posteriorly. Instead the peritreme is normal, joining the stigma anteriorly.

1a   Peritrematal shields fused to ventri-anal shield; usually with a pair of expulsory vesicles posterior on posterolateral to coxae IV; laterodistal elements of pretarsi II-IV considerably longer than associated claws; epigynial shield not fused to ventrianal shield; dorsal seta on the fixed cheliceral digit wedge-shaped .... **Genus** *Neparholaspis*

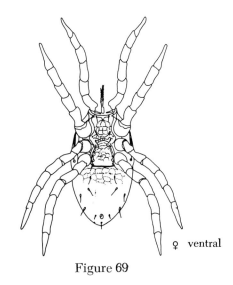

Figure 69

Figure 69 *Neparholaspis evansi* Krantz

Dorsal shield with 30 pairs of simple setae; third pair of sternal setae similar in length to second pair; metasternal shields not fused to sternal or endopodal shields but lying between them; lateral margin of dorsal shield and lateral interscutal membrane without setal pelage; sternal shield with uneven reticulation.

Collection in Oregon from moss, hemlock litter, leafmold, under oak, fir litter.

1b Peritrematal shields not fused to ventrianal shields, expulsory vesicles absent; laterodistal elements of pretarsi II-IV rarely exceeding in length the associated claws, or apparently absent 2

2a Metasternal shields fused to the sternal shield; ventrianal shield with more than four pairs of pre-anal setae ....
.... **Genus** *Calholaspis*

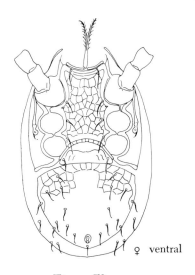

Figure 70

Figure 70 *Calholaspis superbus* Berlese

Ventrianal shield with six pairs of pre-anal setae; tarsus II with stout spurs apically.

Collected in United States and Columbia from humus.

2b Metasternal shields fused to endopodals III or free in the integument; ventrianal shield with three or four pairs of pre-anal setae 3

3a Tectum with two prominent lateral projections, as well as a central extension; ventrianal shield never extending laterally beyond level of coxae IV 4

3b Tectum variously produced, without a pair of prominent lateral projections; ventrianal shield extending laterally beyond level of coxae IV 5

4a Epigynial and ventrianal shields coalesced or fused; movable digit of cheliceral bidentate; metasternal shields free; femur IV without ventral spur ....
.... **Genus** *Neoparholaspulus*

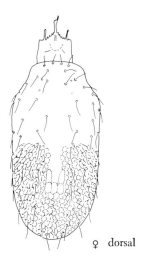

Figure 71

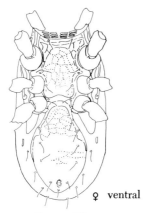

♀ ventral

Figure 72

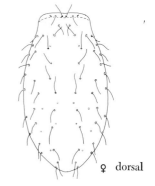

♀ dorsal

Figure 73

Figures 71 and 72 *Neoparholaspulus coales-cens* Krantz

Peritreme extending anteriorly only to a point between coxae I-II, not reflected dorsally; portion of sternal posterior to sternal setae III without ornamentation; dorsal shield with 29 pairs of simple setae.

Collected in Louisiana on soil around sugarcane roots.

4b Epigynial and ventral shields not connected; movable digit of chelicera with two, or more than ten, teeth; metasternal shields free or fused to endopodals III: typically with a distinct ventral spur on femur IV ....
.... **Genus *Parholaspulus***

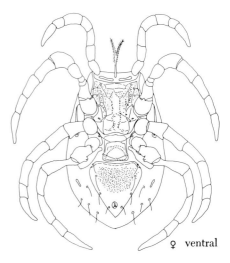

♀ ventral

Figure 74

Figures 73 and 74 *Parholaspulus tragardhi* Krantz

Ventrianal shield triangular; sternal shield with strong reticulation anteriorly; spur on femur IV broad, rounded, sometimes divided distally.

Collected in Oregon from *Pinus contorta* litter, leaf mold, cedar stumps, fir litter, Douglas fir litter.

5a Movable digit of chelicera shorter than corniculus; with well-developed claws on legs I of female .... **Genus *Lattinella***

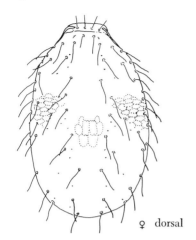

♀ dorsal

Figure 75

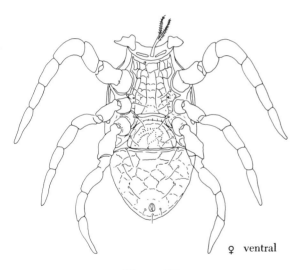

♀ ventral

Figure 76

Figures 75 and 76 *Lattinella capizzii* Krantz

Dorsal shield with 31 pairs of long simple setae, presternal shields not fragmented, movable digit of chelicera shorter than corniculus; with well-developed claws on legs I of female.

Collected in Oregon from Douglas fir litter, moss-straw litter, *Pinus contorta* litter.

5b Movable digit of chelicera as long or longer than corniculus; no claws on legs I of female .... **Genus *Parholaspella***

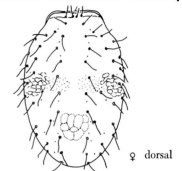

♀ dorsal

Figure 77

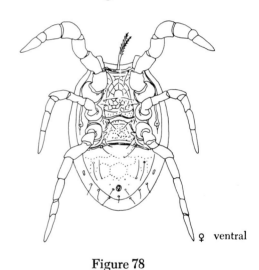

♀ ventral

Figure 78

Figures 77 and 78 *Parholaspella spatulata* Krantz

Movable digit of chelicera with 10-11 teeth; dorsal setae distally spatulate; without spur on femur IV, with three pre-anal setae.

Collected in Oregon from Thuja litter, moss litter, fir tree hole, leaf mold.

Family Dermanyssidae

The classification of the family Dermanyssidae is undergoing a change and not all authors agree as to the placement of the various groups. Evans and Till (1966) have utilized the subfamily classification; such families used by Krantz (1970) are treated as subfamilies.

Within this family, according to Evans and Till (1966), are the Laelapidae, Macronyssidae, Rhinonyssidae, Entonyssidae, Halarachuidae, Ixodorhynchidae and Hystrichonyssidae. This classification would include within a single family, free-living, paraphagic, facultative and obligatory ectoparasites.

The treatment followed in this work is not intended to be an establishment of a system of classification but only to aid a beginning worker to gain insight as to the individual mites that are found in the United States. The Keys established by Camin and Gorirossi (1955) and Krantz (1970) are followed and in no instant is it to be intended that the work of Evans and Till (1966) is not to be followed.

This family is characterized by its members having a stylet-like chelicerae which is modified for piercing the skin of its host. It contains the well-known house mouse mite which is of medical interest due to its ability to transmit certain pathogens to humans. It is a known carrier of *Rickettsia akari* the etiological agent of rickettsialpox. Another species of importance within this family is the chicken mite, *Dermanyssus gallinae,* which can cause painful skin irritation when feeding on people.

1a   Chelicerae attenuated, apical two-thirds or more much narrower than basal portion; chelae minute; anal plate more than half as wide as opisthosoma .... **subfamily Dermanyssinae** 3

1b   Chelicerae of uniform diameter throughout length, if attenuated, chelae obvious 2

2a   Sternal plate not longer than wide, generally much wider than long; anal plate less than half as wide as opisthosoma .... **subfamily Myonyssinae** 4

2b   Sternal plate about twice as long as wide. Parasitic in the auditory meatus of ungulates .... **subfamily Raillietinae** 4

3a   Dorsum with two plates, posterior small, pygidial, anal plate oval, anterior margin rounded .... **Genus *Allodermanyssus***

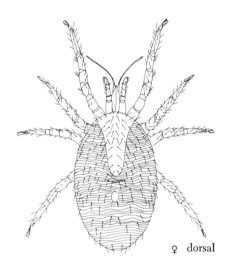

♀ dorsal

Figure 79

Figure 79 *Allodermanyssus sanguineus* (Hirst)

Female with two dorsal plates, posterior plate small, genitoventral plate tapering posteriorly, chelicerae long, whiplike, sternal plate with three pairs of setae and two pairs of slit-like pores, peritreme fused posteriorly with forea of coxa IV, ventral setation sparse, no spurs on coxae or legs.

Collected from many mammal hosts throughout the world but its preferred host is the common house mouse, *Mus musculus.* It has been taken from *Rattus norvegicus, Rattus rattus* and frequently attacks man causing rash. It is the causative agent of rickettsialpox and has been reported to transmit tularemia in Russia. This mite could be expected to occur throughout the United States wherever its preferred host, the house mouse, is found.

3b   Dorsal plate entire. Female sternal plate much wider than long, male holoventral plate slightly expanded behind coxae IV .... Genus *Dermanyssus*

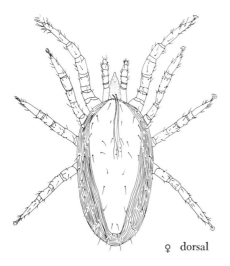

♀ dorsal

Figure 80

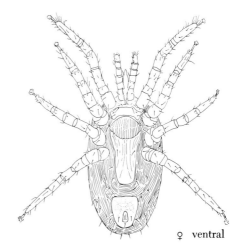

♀ ventral

Figure 81

Figures 80 and 81   *Dermanyssus gallinae* (DeGeer)

Dorsal plate present not divided, tapers posteriorly but not pointed and possesses a truncate posterior margin, sternal setae IV (the metasternals) present, peritreme long, extending at least to middle of coxa II, genitoventral plate with one pair of setae, rounded posterior margin, and plate large, dorsal and ventral body setae sparse, no coxal spurs or spine-like setae on other leg segments.

This is the well-known chicken mite and is associated with domestic fowl, turkey, duck, pigeon, English sparrow, starling, canary and many other birds. It is cosmopolitan in its distribution and could be expected to be found throughout the United States. This species is known to attack man and cause a painful skin irritation. It was believed to be a vector of arthropod-borne virus encephalitides, however, the evidence is yet inconclusive and there are conflicting results.

4a   Sternal plate of female wider than long, anterior margin concave, chelicerae attenuated in both sexes, spermadactyl entirely fused with movable digit of

male, anterior spine of female coxa II prominent .... **Genus** *Myonyssus*

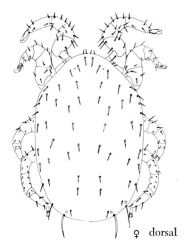

♀ dorsal

Figure 82

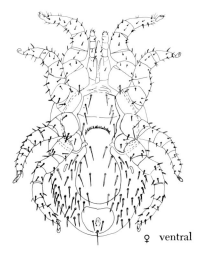

♀ ventral

Figure 83

Figures 82 and 83 *Myonyssus montanus* Furman and Tipton

Female sternal plate only slightly wider than genito-ventral plate, latter plate with not over 15 setae, peritreme extending to anterior one-fourth of coxa II, male with not more than 20 setae on ventral portion of holoventral plate. Collected from *Ochotoma princeps* in Utah.

4b **Sternal plate about twice as long as wide, with a single dorsal plate, restricted to podosoma, multiple rows of deutosternal teeth** .... **Genus** *Raillietia*

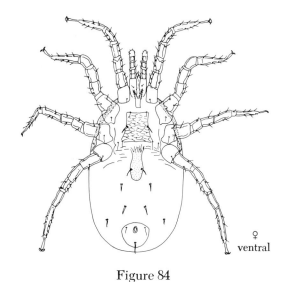

♀ ventral

Figure 84

Figure 84   *Raillietia auris* (Leidy)

With a single dorsal plate, restricted to podosoma, peritremes extending beyond coxa II, tritosternum present, female epigynial plate reduced, male with short sternigenital plate, not extending beyond coxa IV, chelae with modified fused spermatodactyl and movable arm.

Collected from cattle and antelopes where it is found only in the auditory apparatus of cattle. In inhabits not only the outer and middle ear but may also penetrate into the inner ear. This species will probably be found everywhere that cattle are raised.

Family Macronyssidae

This family contains some of the most common parasitic mites associated with man and the domestic rodents and birds. It contains the tropical rat mite, the Northern fowl mite and the tropical fowl mite. Each of these species has caused man to be always aware of the potential danger of these three parasitis species of mites in their ability to transmit various types of pathogenic organisms.

1a   Two dorsal plates present 2

1b   One dorsal plate present 4

2a   Dorsal plates subequal; posterior plate narrower; parasitic primarily on bats and birds 3

2b   Posterior dorsal plate smaller than anterior dorsal plate; posterior without setae; associated with snakes .... **Genus *Ophionyssus***

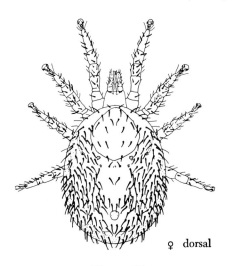

♀ dorsal

Figure 85

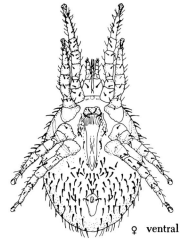

♀ ventral

Figure 86

Figures 85 and 86 *Ophionyssus natricis* (Gervais)

Female with two dorsal plates, anterior plate lemon-shaped, covering most of podosoma, extending posteriorly to level of posterior margins of coxae IV, plate bears nine to eleven pairs of setae, posterior plate is dorsal to anal plate, smaller, setae absent on plate, chelicerae without ventral palpal spur, two pairs of setae on sternal plate, genitoventral plate pointed posteriorly.

Collected from all types of snakes in zoos. It is a dangerous blood-sucking parasite of captive snakes. This species is called the snake mite and it is the vector of a bacterium *Pseudomonas hydrophilus* which causes a fatal disease called hemorrhagic septicemia in captive reptiles. It also transmits a haemogregarine of snakes. It has been recorded from other animals such as rats and people, some snakes recorded as hosts are *Coluber florulentus, Coluber karelini, Python reticulatus* to name only a few. It has been recorded in many zoos and pet stores throughout the United States and the world.

3a Sternal plate of female twice as wide as long; posterior margin of sternal plate strongly sclerotized, chelicerae not attenuated .... **Genus** *Steatonyssus*

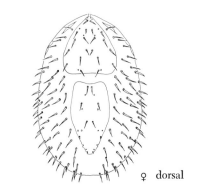

♀ dorsal

Figure 87

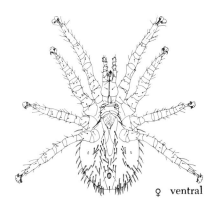

♀ ventral

Figure 88

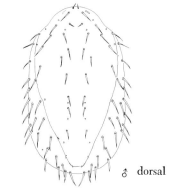

♂ dorsal

Figure 89

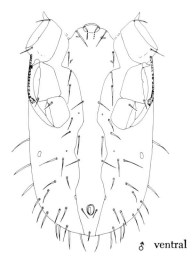

♂ ventral

Figure 90

Figures 87, 88, 89, and 90 *Steatonyssus antrozoi*

3b Sternal plate of female very narrow, arched, without a sclerotized posterior margin. Sternal setae I much smaller than sterna setae II, cheilicerae long, attenuated .... **Genus** *Pellonyssus*

Female anterior dorsal shield with a straight posterior border, setae at posterior of opisthosoma not enlarged over ventral body setae, peritremal plate elongate, running from anterior level of coxa II, curving beneath coxa IV and coalescing with endopodal plate, sternal shield ten times as wide as long. Protonymph with two small dorsal platelets in addition to large anterior dorsal plate and smaller pygidial plate.

Collected from *Passer domesticus, Mimus polyglottis,* in Maryland, Florida and Texas.

4a Coxal spurs present at least on some coxae; palpal trochanter of female lacking spur; epigynial plate rounded posteriorly. Male ventral plate undivided behind coxa four ...... **Genus *Patrinyssus***

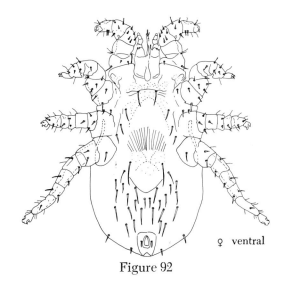

♀ ventral

Figure 92

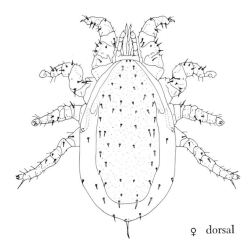

♀ dorsal

Figure 91

Figures 91 and 92   *Patrinyssus hubbardi* Jameson

Female without spurs on basal palp segment, genitoventral plate mended posteriorly, some coxal spurs setigerous, male ventral plate divided behind fourth coxae.

4b Coxal spurs absent, coxal setae present and may be thorn-like in structure 5

5a Epigynial plate narrow, pointed; dorsal plate broad anteriorly, tapering posteriorly; spur present on ventral surface of palpal trochanter, parasitic only on birds and mammals ....
.... **Genus *Ornithonyssus***

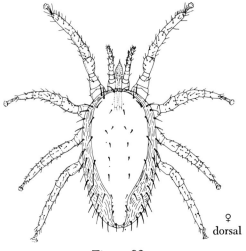

Figure 93

This mite is known as the Northern fowl mite and is a very serious pest of domestic fowl and wild birds. It apparently will not attack mammals readily, however, it has been recorded as feeding on man and some domestic rodents, but does not prefer mammals as a steady source of food. It is considered to be capable of transmitting fowl pox and is naturally infected with western equine encephalitis.

Some of its hosts include blackbirds, chickens, kingbirds, meadowlarks, pigeons, grackles, robins, sparrows, starlings, to name only a few. This species is found throughout temperate regions of the world as a parasite of domestic fowl.

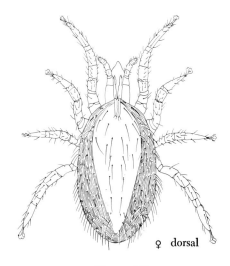

♀ dorsal

Figure 95

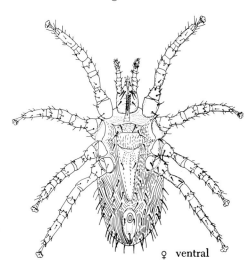

♀ ventral

Figure 94

Figures 93 and 94 *Ornithonyssus sylviarum* (Canestrini and Franzago)

The sternal plate with two pairs of setae, the third posterior pair being located posterior to the plate, although occasionally they may touch it, dorsal plate setae equal in size to those setae on the unprotected dorsal integument, sternal plate of female reduced and with only two pairs of setae.

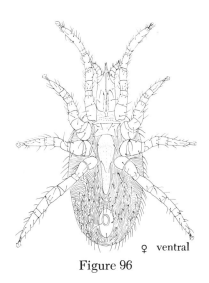

Figure 96

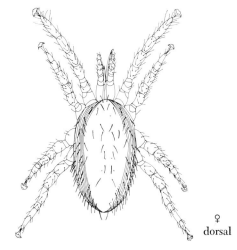

Figure 97

Figures 95 and 96   *Ornithonyssus bacoti* (Hirst)

Female with a single dorsal plate, relatively narrow, several pairs of long setae on dorsal plate, chelicerae toothless without pilis dentilis, distal segment of palp with a ventral spurlike process, sternal plate with three pairs of setae, genitoventral plate narrows gradually to the rear, making its posterior end appear pointed, genitoventral setae absent, three anal setae present, peritreme fused posteriorly with the fovea of coxa IV, coxa II with a sharply pointed anterodorsal spur.

This is the well-known tropical rat mite and has a cosmopolitan distribution. It is associated with domestic and wild rats in both tropical and temperate areas of the world. It will readily bite people living in buildings infested with rats. It may cause painful dermatitis. It has been reported from a wide range of hosts, such as domestic cats, rats, skunks, squirrels, house mice, field mice, opossum, raccoons, shrews, white mice and poultry to name only a few. It has been recorded throughout the world in association with domestic rats.

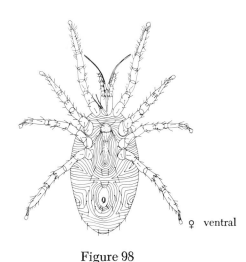

Figure 98

Figures 97 and 98   *Ornithonyssus bursa* (Berlese)

Setation of dorsal plate smaller than those on the dorsal integument, sternal plate has three pairs of setae, anterior pair being in the anterior margin of the plate, female with dorsal plate broad.

This is the well-known tropical fowl mite. As a serious pest of domestic poultry it is only surpassed by *Dermanyssus gallinae*, the chicken mite. This mite is restricted to the warm and tropical regions. It can cause discomfort to man but can only live for a short period off of its natural bird host. Records of this species from northern sections of the United States are probably due to mis-identification of the other species of *Orni-thonyssus*.

Collected from chickens, common sparrow, ducks, English sparrow, starling, king-bird, meadowlark, pigeons, setting hens, turkeys to list only a few.

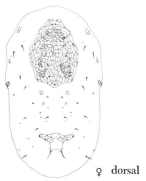

♀ dorsal

Figure 99

Key to the Three Common Species of Ornithonyssus

a   Female with dorsal plate broad, dorsal plate setae short, reaching about half-way to bases of setae of next row, on birds .... **b**

aa   Female with dorsal plate narrow, tapering rapidly posteriorly, dorsal plate setae long, reaching to or past bases of setae of next row giving mite a "hairy" appearance, on rats .... *Ornithonyssus bacoti*

b   Sternal plate of female with three pairs of setae .... *Ornithonyssus bursa*

bb   Sternal plate of female reduced and with only two pairs of setae .... *Ornithonyssus sylviarum*

5b   Epigynial plate rounded, dorsal plate parallel-sided, covering most of dorsal surface **6**

6a   Legs I and II thick and heavy (figs. 99 & 100) .... **Genus** *Radfordiella*

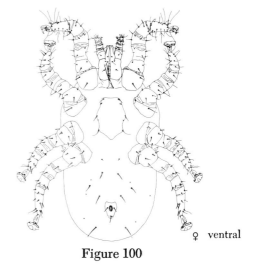

♀ ventral

Figure 100

6b   Leg I and II normal, not thick and heavy **7**

7a   Coxae without tubercules; ventral plates of male fused into a single holoventral plate (figs 101 & 102) .... **Genus** *Macronyssus*

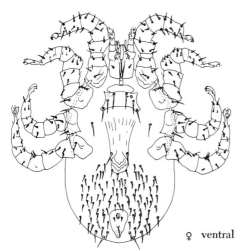

♀ ventral

Figure 101

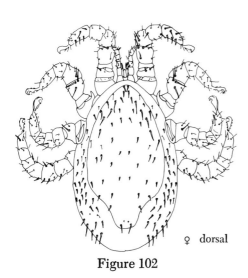

♀ dorsal

Figure 102

7b Coxae II and III with tubercles, ventral plate of male divided into sternogenital and a ventroanal plate; parasitic on bats only .... Genus *Ichoronyssus*

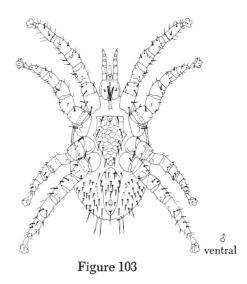

♂ ventral

Figure 103

Figure 103 *Ichoronyssus quadridentatus* Strandtmann and Hunt

Female dorsal plate parallel-sided, narrow, gradually anteriorly and abruptly posteriorly, sternal plate twice as wide as long with three pairs of setae, male dorsal shield widest in shoulder regions and tapers evenly to a broad heavily sclerotized four-toothed plate at posterior end.

Collected from *Eptesicus fuscus, Myotis lucifugus, Myotis austroriparius* in Georgia and Florida.

Family Phytoseiidae

The members of this family are free-living, predaceous, mesostigmatid mites and constitute a distinct group from closely related families. The number of species known are increasing and will continue to do so as more attention is given to those animals that can be used to obtain a balance between the destructive insects and mites and their natural enemies. Following the work of Chant (1965) in which are outlined the general concepts of the family, two subfamilies of the Phytoseiidae are recognized, the Otopheidomeninae and the Phytoseinae.

1a   Anus in terminal position; fixed digit of chelicera of female reduced or absent .... Subfamily Otopheidomeninae 2

1b   Anus in ventral position; fixed digit of chelicera normal .... Subfamily Phytoseiinae 4

The Otopheidomeninae is now considered to be a family consisting of the ear-sparing mites of moths.

2a   Metasternal setae absent in deutonymph and adults .... Genus *Otopheidomenis*

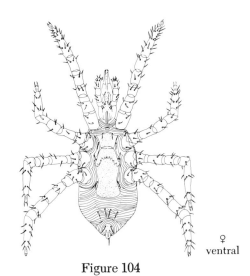

♀ ventral

Figure 104

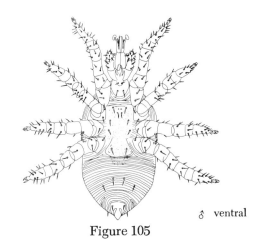

♂ ventral

Figure 105

Figures 104 and 105   *Otopheidomenis zalelestis* Treat

The dorsal shield incised laterally; two pairs of sublateral setae, r5 and R1 on lateral integument; sternal shield with two pairs of setae.

Collected from moths of the genus *Zale* in New York, New Jersey, Georgia, Mississippi, Alabama, and Florida.

2b   Metasternal setae present in deutonymph and adults 3

3a   Pedipalp with 24 setae, palpal genu without setae or at most with a single setae; genua II to IV with total of 11 setae .... Genus *Hemipteroseius*
The two species contained within this genus are found only from insects of the order Hemiptera from Nigeria and India.

3b   Pedipalp with more than 24 setae, palpal genu with five or six setae; genu II to IV with 19 to 21 setae; palpal femur with macrosetae; genital shield subdivided into subequal anterior and posterior shields .... Genus *Treatia*

Treatia phytoseioides (Baker & Johnston)

Dorsal shield divided into subequal anterior and posterior portions with 15 pairs of setae; two pairs of sublateral setae; r5 on dorsal shield, and R1 on lateral integument; sternal shield with three pairs of setae.

Collected from an insect of the order Hemiptera from Florida.

4a **Dorsal shield laterally divided into sub-equal shields** .... **Genus** *Macroseius*

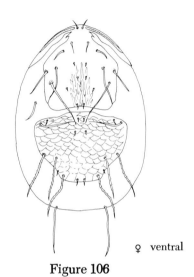

♀ ventral

Figure 106

Figure 106 *Macroseius biscutatus* Chant, Denmark and Baker

Dorsal shield subdivided into anterior and posterior shields with S1 in the anterior dorsal shield and S2 on the interscutal membrane near the posterior margin of anterior shield.

Collected in Florida from *Sarracenia* sp., Alachua Co., Florida.

4b **Dorsal shield entire** **5**

5a **Dorsal shield with five or six pairs of prolateral setae, one being added in the second molt** **6**

5b **Dorsal shield with four pairs of prolateral setae** **8**

6a **Anterior sublateral seta (r5) on dorsal shield; posterior sublateral seta on lateral integument or absent; dorsal shield with setae serrated when thickened; genu IV with seven setae** .... **Genus** *Phytoseius*

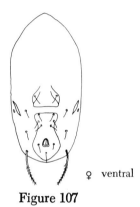

♀ ventral

Figure 107

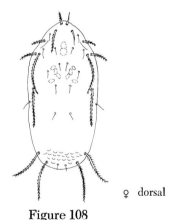

♀ dorsal

Figure 108

Figures 107 and 108 *Phytoseius plumifer* (Canestrini and Fanzago)

Dorsal shield with 16 pairs of setae, seven in lateral rows. Leg IV with one macroseta.

Collected from California on grape

vines, also known to occur throughout Italy in association with grapes.

6b Anterior sublateral seta on lateral integument genu II with seven or eight setae, genu III with six or seven setae 7

7a Genu III with six setae; hypostome, palps, and chelicerae greatly elongated, the hypostome and palp together as long as leg I .... Genus *Gigagnathus*

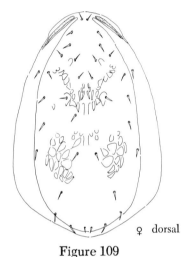

♀ dorsal

Figure 109

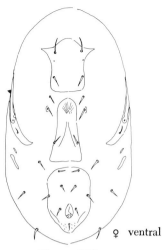

♀ ventral

Figure 110

Figures 109 and 110 *Gigagnathus extendus* Chant

Hypostome narrow, greatly elongate, chelicerae very elongate, fixed digit with one tooth, movable digit without teeth; palps greatly elongate. Dorsal shield with 17 pairs of setae, nine in the lateral, two in the mediolateral and six in the dorsocentral rows.

Collected on *Chamaecyparis* sp. from Bermuda grass at Boston, Massachusetts.

7b Genu III with seven setae; hypostome, palps, and chelicerae not greatly elongated, the hypostome and palp together much shorter than leg I .... Genus *Typhlodromus*

The genus *Typhlodromus* has, at the least, 27 species recorded from the United States. This number is only an estimate and is dependent on which classification a worker follows and how thorough the literature is searched. The recent work by Chant (1965) should be utilized; however, all workers do not follow this classification and a student working with this group should consult the work of Muma

(1963) concerning the genus *Galendromus* Muma.

8a   Anterior sublateral seta (r5) on the dorsal shield in dorsal position; some setae on dorsal shield greatly flattened, serrated .... Genus *Platyseiella*

8b   Anterior sublateral seta on the lateral integument, or on lateral sclerotized extension of dorsal shield in lateral position; some setae on dorsal shield may be serrated but never greatly flattened 9

9a   Lateral integument sclerotized so that setae r5 and R1, though in usual lateral position, are on lateroventral extension of dorsal shield .... Genus *Iphiseius*

9b   Lateral integument not sclerotized so that setae r5 and R1 are located on lateroventral extension of dorsal shield 10

10a Metapodal plates single, large, triangular; genital and ventrianal shields very broad, the former punctate; genu II with seven setae, genu III with six setae .... Genus *Paraamblyseius* Muma

10b Metapodal plates paired, slender, elongate; genital and ventrianal shields narrower, the former not punctate; genu II with seven or eight setae, genu III with seven setae 11

11a Body large, globular, reddish color when alive; dorsal shield small .... Genus *Phytoseiulus*

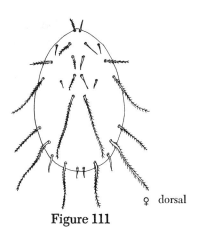

Figure 111

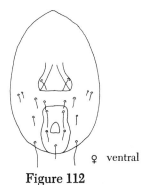

Figure 112

Figures 111 and 112 *Phytoseiulus macropilis* (Banks)

Dorsal shield with 14 pairs of setae, seven in the lateral rows. Ventrianal shield almost square, with a single pair of pre-anal setae.

Collected from water hyacinth, Eustis, Florida, also taken from Hawaii, British West Indies, Canary Islands, and the Panama Canal Zone.

11b Body compressed dorsally, hyaline or brownish when alive, dorsal shield large (fig. 113) .... Genus *Amblyseius*

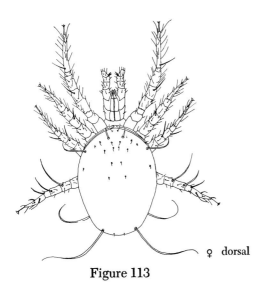

♀ dorsal

Figure 113

The genus *Amblyseius* is a very large genus according to the classification of Chant (1965). Many of the genera established by Muma from Florida are synonymized by Chant and placed in *Amblyseius*. Any student attempting to work on members of this genus found in the United States should make a thorough study of the works of Chant, 1959, 1965; Muma, 1961, 1965; DeLeon, 1963; Schuster and Pritchard, 1963; and Pritchard and Baker, 1962.

Family Ascidae

In the past this family has undergone several changes in classification. The members now included within this family were contained in two families, the Blattisociidae and the Aceoseiidae. The work of Lindquist and Evans (1965) is followed and the use of the family name Ascidae is here utilized on the basis that it is the oldest valid family-group name for this group of mites.

There are approximately 350 species of Ascidae distributed throughout the world. Three subfamilies are utilized, Arctoseinae, Platyseiinae and Ascinae. The latter subfamily is broken down into three tribes, Ascini, Blattisocini and Melicharini.

The members of this family are to be found in diversified habitats. The Arctoseinae are associated with litter and soil humus and are free-living. They are believed to feed on small soil organisms such as nematodes. Some of the members of the subfamily Platseiinae are associated with water and are found in marshes, stream mosses, spray zones of waterfalls and damp sod. Other species of this subfamily are found in dryer conditions such as grasslands and forest litter. Members of this group are considered to be predators of small invertebrates. The Ascinae make up a very diversified group which includes free-living predators of salt marshes and sea shores, nests of vertebrates and arthropods, adult bumble bees and nests, the nares of hummingbirds, flowers and plant foliage, stored foods, injured and dying trees, bracket fungi, various insects associated with the spiracula atria, and millipedes. Their food preferences are as varied as their habits.

1a   Dorsal shield of deutonymph and adult united; lateral incisions of deutonymph retained or obliterated on adults 2

1b   Dorsal shield of deutonymph and adult completely divided into two separate shields 5

2a Posterior dorsal shield in both sexes with a pair of conspicuous setae-bearing horns posterolaterally .... Genus *Asca*

2b Posterior dorsal shield without conspicuous setae-bearing horns posterolaterally 3

3a Minimum number of setae on genua I, II, III, IV: 12-11-8-9, on tibiae: 12-10-8-10; opisthonotum usually with five pairs of lateral setae; dorsal shield holotrichous, genu and tibia each with 13 setae .... Genus *Leioseius*

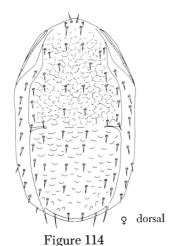

♀ dorsal

Figure 114

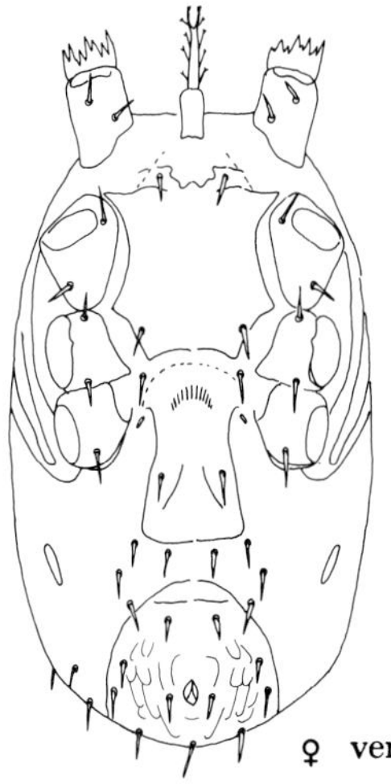

♀ ventral

Figure 115

Figures 114 and 115 *Leioseius californicus* Chant

The ventrianal shield with three pairs of preanal setae; tectum bispinate. Collected from litter of laurel plants in California.

3b Maximum number of setae on genua I, II, III, IV: 12-10-8-7, on tibia: 12-9-7-7; Opisthonotum usually with 4 pairs of lateral setae 4

4a Tarsi II-IV with at least one dorsolateral subapical setae very slender and elongated; palptarsus with macroseta; vertex of dorsal shield strongly arched downward, concealing vertical setae; peritremes long, their anterior extremities sharply recurved posteroventrally .... Genus *Iphidozercon*

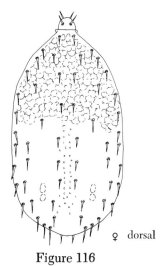

Figure 116

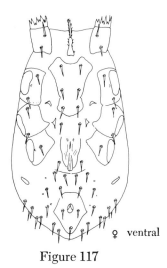

♀ ventral

Figure 117

Figures 116 and 117 *Iphidozercon californicus* Chant

Dorsal shield curving ventrally along lateral margin, marginal setae difficult to observe.

4b **Tarsi II-IV with neither dorsolateral subapical setae slender and elongated; vertex of dorsal shield not strongly arched, vertical setae visible from above;**

peritremes short, their anterior extremities not recurved .... **Genus *Arctoseius***

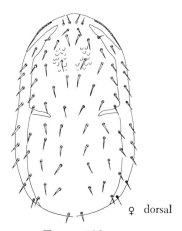

♀ dorsal

Figure 118

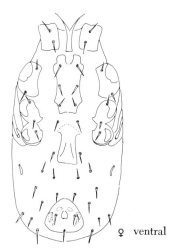

♀ ventral

Figure 119

Figures 118 and 119   *Arctoseius cetratus* (Sellnick)

Tectum bispinate in female; peritreme not extending anteriorly of coxa II, sternal shield with three pairs of setae, smooth.

Collected from Tennessee, Texas, and Oregon, from soil and under cow manure.

5a Fixed chela with setiform pilus dentilis; movable chela without ventral mucro; peritrematal shield of adults broadly connected posteriorly to exopodal plate curving behind coxa IV 6

5b Fixed chela, with membranous lobe in place of pilus dentilis, movable chela usually with ventral mucro near base, peritrematal shield of adults free or narrowly attached posteriorly to exopodal plate beside coxa IV 9

6a Corniculi slender, contiguous; rows of deutosternal denticles narrow, each with few denticles, tectum rounded, smooth .... Genus *Blattisocius*

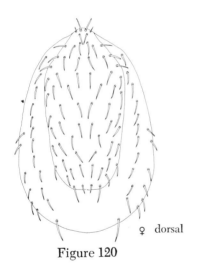

Figure 120

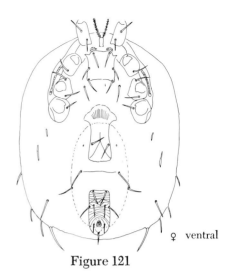

Figure 121

Figures 120 and 121   *Blattisocius tarsalis* (Berlese)

Peritreme extending to level of posterior margin of coxa II; fixed digit of chelicera very short, without teeth.

Collected from Oregon, California, New York in stored wheat, barley, roach cultures. This species is recorded as parasitic on moths in stored food products. It feds on the eggs and larvae of *Ephestia* and *Sitotroga*, both grain moths.

6b Corniculi stout, well separated; rows of deutosternal denticles narrow or wider, each with few to many denticles, fixed chela well-developed 7

7a Rows of deutosternal denticles narrow, each with two to four denticles; adults with all marginal setae on edge of dorsal shield .... Genus *Neojordensia*

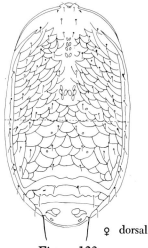

♀ dorsal

Figure 122

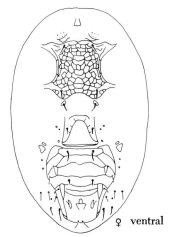

♀ ventral

Figure 123

Figures 122 and 123   *Neojordensia tennes-seensis* DeLeon

Dorsal shield setae short; ventrianal shield longer than wide; five pairs of interscutal setae; peritrematal shield fused with exopodal shield in region of coxa IV; legs without macrosetae.

Collected in Tennessee from a punky hardwood log.

7b Rows of deutosternal denticles moderately wide, each with five to many denticles; adults with three-nine pairs of marginal setae on lateral membrane 8

8a Female with 11 pairs of setae on anterior region of dorsal shield and seven to 10 pairs on posterior region; genital shield rounded posteriorly; cervix of spermatheca not sclerotized ....
.... Genus *Acerodromus*

8b Female with 12-23 pairs of setae on anterior region of dorsal shield, and 10-15 pairs on posterior region, genital shield usually truncate posteriorly; cervix of spermatheca well sclerotized; movable chela usually tridentate; genua II and III usually with 11 or nine setae respectively .... Genus *Lasioseius*

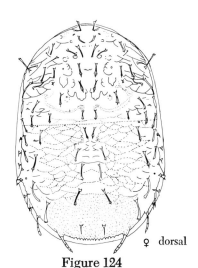

♀ dorsal

Figure 124

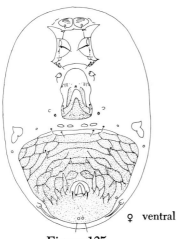

♀ ventral

Figure 125

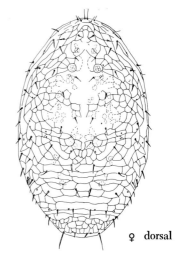

♀ dorsal

Figure 126

Figures 124 and 125 *Lasioseius spectabilis* DeLeon

Dorsal shield with 15 pairs of setae; flanged; the metapodalia fused with each other; chelicerae fixed digit with five-six large teeth proximally. Peritremata shield not fused with exopodal in region of coxa IV.

Collected in Tennessee, in association with fruiting body of fungus *Peniphora gigantea*, on log of *Pinus strobus*, and under bark of dead *Liriodendron tulipifera*.

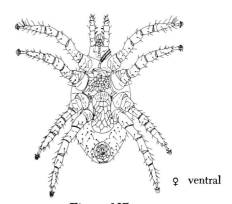

♀ ventral

Figure 127

9a   Adults with seven-13 pairs of marginal (R) setae on lateral membrane; female with maximum of 15 pair of setae on posterior region of dorsal shield .... Genus *Melichares*

9b   Adults with zero-three pairs of posterior marginal (R) setae on ventrolateral membrane with all others on edge of dorsal shield; female with 18 to 22 pairs of setae on posterior region of dorsal shield .... Genus *Proctolaelaps*

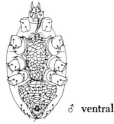

♂ ventral

Figure 128

Figures 126, 127, and 128 *Proctolaelaps dendroctoni* Lindquist and Hunter

Dorsal shield with 42 pairs of setae; three most posterior pairs of marginals on membrane behind rounded posterolateral corners of dorsal shield. Preendopodal plates absent, sternal shield well defined.

Collected in Texas, Louisiana and Georgia, from galleries of *Dendroctonus frontalis*, *Ips avulsus*, both bark beetles.

Family Digamasellidae

These are mesostigmatid mites known to be found in association with bark beetles and their galleries. Some species occur under bark in forest litter, soil, dung, salt marshes and even the nests of ants. This family in its present state contains only two genera, *Digamasellus* and *Longoseius*, the latter is a very large genus containing at least 80 some species distributed throughout the world.

1a   Body very long and narrow, poorly sclerotized, anterior dorsal shield with 12 pairs of simple setae, posterior shield with 14 pairs, all short ....
.... **Genus *Longoseius***

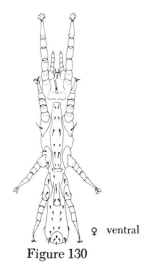

♀ ventral

Figure 130

Figures 129 and 130 *Longoseius cuniculus* Chant

Idiosoma very long, narrow, with two dorsal shields, weakly sclerotized, reduced number of setae; marginal series of setae almost absent, sternal shield long, narrow, not fused with endopodal plates, with three pairs of setae, three pairs of minute pores.

Collected in Maine, Louisiana, Mississippi from the sawger beetle, *Monochamus notatus* (Coleoptera-Cerambycidae), *Ips calligraphus* from loblolly pine, *Dendroctonus frontalis*.

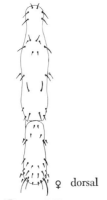

♀ dorsal

Figure 129

1b   Body normal with usually more than 12 pairs of anterior dorsal shield setae;

male with a pair of triangular plates between coxae IV bearing genital setae ....
.... **Genus *Digamasellus***

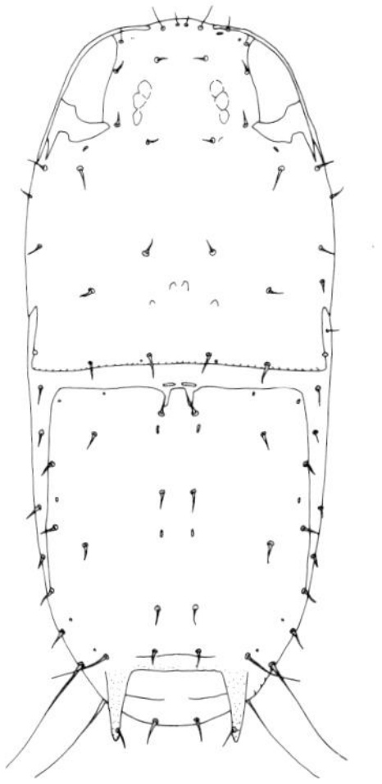

of setae V-6. Prominent horns on posterior
dorsal shield.

Collected in Delaware and Ohio from
elm bark beetle, *Scolytus multistriatus* galleries.

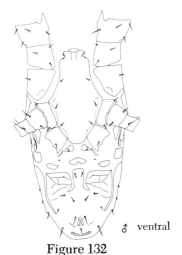

Figure 131

♂ dorsal

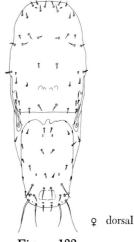

♀ dorsal

Figure 133

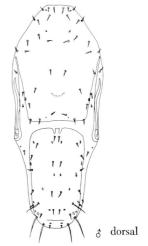

Figure 132

♂ ventral

Figures 131 and 132 *Digamasellus moseri*
Hurlbutt

Female ventrianal with only two pairs of
setae; setae V-3 absent, incision in ventrianal
of male extending posteriorly nearly to level

♂ dorsal

Figure 134

Figures 133 and 134 *Digamasellus brachypoda* Hurlbutt

Genu III with 7 setae; tibia IV of male with a posteriorly projecting spine, posterior margin of idosoma convex, peritreme reduced anteriorly, tarsus I and tibia I short.

Collected in Louisiana from vacated galleries of *Dendroctonus frontalis* and inner bark of loblolly pine.

Family Zerconidae

This family occurs with other mites in the upper soil layer of forests. Many of the known species are found with moss, grass roots from alpine meadows, juniper and hardwood litter. The work of Halaskova (1969) and Sellnick (1958) have been followed in the establishment of the North American fauna.

1a   **Peritrematal shield seta two (p2) short, not plumose** .... **Genus *Prozercon* Sellnick**

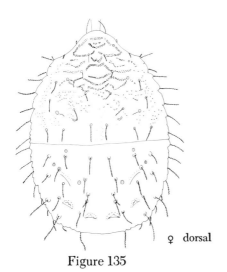

♀ dorsal

Figure 135

Figure 135   *Prozercon praecipuus* Sellnick

Notocephale with rows of small knobs, anterior region of plate with a network of indistinct ridges, posterior half only lines and pits, all hairs barbed except for i:5 which is smooth, notogaster with no visible lines or pits, pore Po_3 lies outward from the line J:2-J:3, nearer J:2 than J:3, all hairs on notogaster barbed, peritremal shield end in a line with posterior border of coxa IV, near outer border of ventral side of notogaster a narrow plate tapering toward posterior end which is fused with ventrianal plate.

Collected on redwood or laurel from California.

1b   **Peritrematal shield seta p2 long, plumose** **2**

2a   **Large ventrianal pores situated posterolaterally of adanal setae, seta p2 located in anterior third section of peritrematal shield, slit between peritrematal shield and notocephale shaped as a broad wedge, anterior margin of notocephale not strongly sloped on under side** .... **Genus *Zercon* Koch**

2b   **Large ventrianal pores located anterolaterally from the insertion of adanal setae, termination of peritrematal shield in a traverse direction, pl located at same level as setae S_3, seta r_3** .... **Genus *Amerozercon* Halašková**

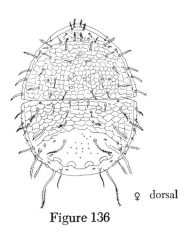

Figure 136

Figure 136 *Zercon farrieri* Halašková

Dorsocentral setae series (J) with only five setae, mediolateral seta series (Z) with only four setae, anterior edge of ventrianal with two pairs of setae, pores (Po$_2$) are located anteromedially from the insertions of setae Z$_3$, pores (Po$_3$) anteromedially from insertions of setae Z$_4$. Setae J$_5$ and Z$_2$ absent, marginal setae series (R) with R$_1$ long, feathered; other R setae shorter than R$_1$ not feathered, insertions of setae J$_6$ and Z$_5$ set close to each other.

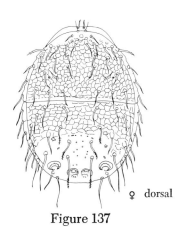

Figure 137

Figure 137 *Zercon insolitus* Halašková

Podonotum reticulated in middle, tile-shaped on sides, middle depressions of reticulated podonotum strikingly large, with axes converging posteriorly, pores Po$_2$ located mediad of a line connecting insertions of setae Z$_1$-Z$_3$, Po$_3$ within line connecting Z$_3$-Z$_4$, setae of podonotum long, smooth, opisthonotum reticulated anteriorly, sparsely pitted posteriorly, setae of opisthonotum long, barbed, insertions of setae J$_6$-Z$_5$ close to each other.

Collected in hardwood litter from North Carolina.

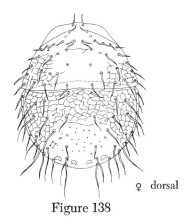

Figure 138

Figure 138 *Zercon comatus* Halašková

Podonotum smooth, reticular sculpture only found in back corners, setae of podonotum smooth or finely pilose, long, pores Po$_3$ are shifted forward, situated mediad of a line connecting the insertion of setae Z$_2$-Z$_3$ and on a line connecting pores Po$_2$, setae of opisthonotum long and smooth, reticular sculpture conspicuous, sclerotized, lateral depressions larger than interior, their axis parallel to body axis.

Collected in loblolly pine litter from South Carolina.

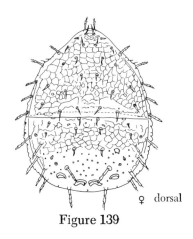

♀ dorsal

Figure 139

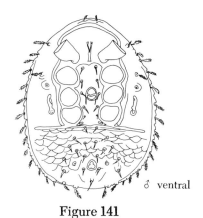

♂ ventral

Figure 141

Figure 139 *Zercon carolinensis* Halašková

Podonotal marginal setae long, feathered, opisthonotum marginal setae short, thorn-like, smooth, reticular sculpture conspicuous well-developed on podonotal shield and anterior portion of opisthonotal shield.

Figures 140 and 141 *Amerozercon suspiciosus* Halašková

With the structures of the genus, only a single male specimen has been found in loblolly pine litter from South Carolina.

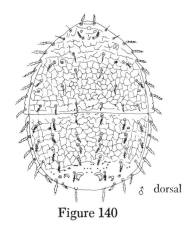

♂ dorsal

Figure 140

Family Laelapidae

This is an extremely complex group of mites. The family classification has in recent years been based on those members that are parasites of animals associated with people. How-ever, if the family as a whole is considered it exhibits a very wide variety of forms. The literature contains a vast amount of articles dealing with those parasitic groups which may prove to be vectors of diseases that affect man or his domestic animals. Little information is known regarding the free-living species of this family, nor has there been an attempt to relate the two groups within the family.

1a **Dorsal and ventral shields and inter-scutal membranes densely clothed with setae; dorsal shield in the nymphal stages, one and two with undivided dorsal shield. Mostly found in the nests of birds and mammals .... Subfamily Haemogamasine 2**

1b **Dorsal and ventral shields with a moderate covering of setae, nymphal stages, one and two, with divided dorsal shield .... Subfamily Laelapinae 3**

This subfamily is well-known because of its connection with rodents and its involvement with transmission of certain disease organisms. They are classified as parasites of rodents and have been proven to be reservoirs of such diseases as plague, tularemia, relapsing fever and some rickettsial infections. The life cycle of members of this family has been studied and it is known that the females are larviparo-cus; the larva does not feed, like all larvae, it is six-legged in this stage. The protonymph stage is the first feeding form and will take a blood meal from its host within a few hours.

Within four-eight days the protonymph will molt into the third stage, second nymphal stage, called the deutonymph. This stage, like the protonymph, will also feed on blood and will in a short period of time molt for the final time into an adult. The adult will also feed on the blood of its host.

Males within this family transfer the sperm by using their chelae. A clear bubble (the spermatophore) forms in the males genital aperature, when completed it is removed by the chelae and inserted into the female atrium. The bubble spermatophore bursts and forces the sperm into the female genital tract.

The group is characterized as having a modified palpal claw with a two-tined palpal apotele. Dorsal plate covering most of the idiosoma in all but the protonymph. This latter stage has two dorsal plates and several platelets between these two plates. Peritremes extending past the coxa of leg III (reduced in protonymph). Reproductive epigynial plate rounded posteriorly not ending in a tapering point.

2a **Tectum elongated, tongue-like, with a fimbriated margin, none of the setae of leg II noticeably different from those of legs III and IV .... Genus Haemogamasus**

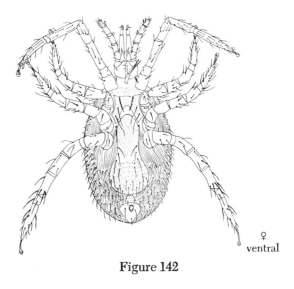

♀ ventral

Figure 142

Figure 142 *Haemogamasus pontiger* (Berlese)

Female with some barbed setae chelae with teeth, setae at posterior body margin only slightly enlarged, fixed cheta with only one tooth, posterior margin of sternal shield invaginated to a level midway between anterior and median pairs of sternal setae. Male with anal region incorporated in ventral shield, with some barbed setae, fixed chela toothless, anterior pair of maxillary setae smooth, accessory sternal setae lacking, shorter branch of movable chela undivided. Nymphs without accessory setae on ventral shield, chelae with teeth, with some barbed setae, fixed chela with a single tooth, base of almost every dorsal body setae with a posteriorly directed thorn-like process, anterior pair of maxillary setae smooth.

Collected from the nest of *Tamiasciurus fremonti*, associated with clothes moth, and on a rug, from Colorado, New York, Oregon, Utah and Nevada.

2b   **Tectum with a truncate, fimbriated margin. Legs II with some heavy spines and spurs .... Genus *Ischyropoda***

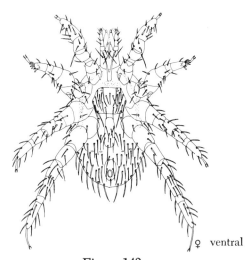

Figure 143

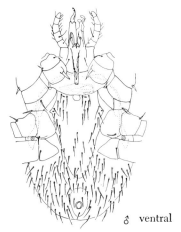

♂ ventral

Figure 144

Figures 143 and 144   *Ischyropoda spiniger* Keegan

Female with movable chelae toothless, all setae on dorsal shield smooth, leg I with a stout posteriorly directed coxal spur, genu of leg II with a similar spur. Male with a separate anal shield, coxa I and genu II each with a large ventral spur, ventral shield widest at level of coxae II. Nymphs with accessory setae on ventral shield, coxa I with a large, posteriorly directed ventral spur, genu II with a similar spur, movable chela toothless.

Collected from *Peragnathus penicillatus* and *Perognathus spinatus* in California and Nevada.

2c   **Metapodal plates large and prominent, triangular in shape. Anal plate triangular, broader than long, bearing only three setae .... Genus *Eulaelaps***

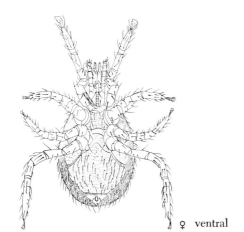

Figure 145

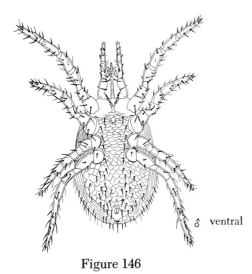

Figure 146

2d **Sternal plate poorly developed bearing only sternal setae II** ....
.... **Genus** *Brevisterna*

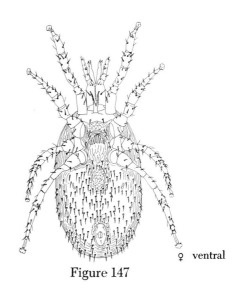

Figure 147

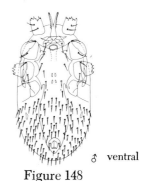

Figure 148

Figures 145 and 146   *Eulaelaps stabularis* (Koch)

Female with large triangular metropodal plates, a broad, triangular anal plate and a greatly expanded epigynial plate. Male with holoventral plate expanded behind coxae IV. Both sexes with peritremalia broad and short, bearing a large stigma-like pore near the apex, chelae dentate, well-sclerotized.

Figures 147 and 148   *Brevisterna morlani* Strandtmann and Allred

Female sternal plate bearing only the middle pair of usual sternal setae, no accessory setae, with a pygidial plate, ventral plate with 15 to 23 accessory setae, anal plate nearly always with one or two accessory setae. Male holoventral plate with 35 to 50 accessory setae between the genital and paranal setae, no barbed setae in any stage.

Collected from *Cynomys gunnesoni, Neotoma albigula, Neotoma micropus, Peromyscus leucopus, Peromyscus maniculatus, Peromyscus truei, Sylvilagus auduboni, Dipodomys ordii* in Colorado, New Mexico and Texas.

Work has been done on this species regarding its importance in the epidemiology of rodent diseases, and rodent-borne diseases.

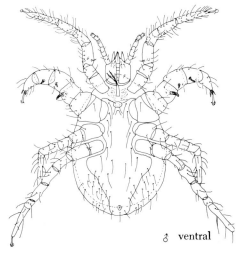

♂ ventral

Figure 150

Figures 149 and 150   *Androlaelaps grandiculatus*

3a   **Femur II and genu II with large thumb-like or conical spurs** ....
.... **Genus** *Androlaelaps*

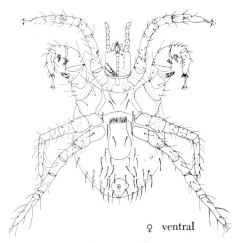

♀ ventral

Figure 149

3b   **Femur II and genu II without large thumb-like or conical spurs** 4

4a   **With two very long spine-like setae dorsally on femur I, epigynial plate with four pairs of setae, peritremalia short, not encircling coxa IV** .... **Genus** *Laelaps*

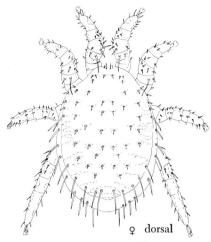

♀ dorsal

Figure 151

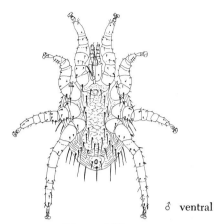

♂ ventral

Figure 153

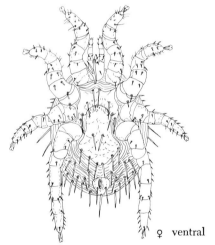

♀ ventral

Figure 152

Figures 151, 152 and 153 *Laelaps kochi Oudeman*

Posterior border of sternal plate invaginated to level at least midway between first and second pairs of sternal setae. This strongly covered caudal margin of the sternal plate along with the separate anal plate of the male and the spiniform dorsal setae distinguishes this species.

4b   Without two long spine-like setae dorsally on femur I 5

5a   Epigynial plate with three or more setae. Coxae I, II, and III with heavy, thorn-like setae, female chelae with movable arm, is sclerotized, toothed and subtended by two very long setae, the other arm is nonsclerotized, slender, and also movable ....
.... **Genus *Steptolaelaps* Furman**

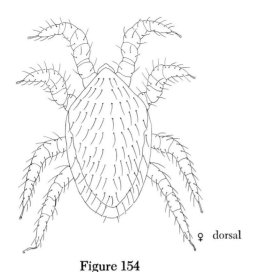

♀ dorsal

Figure 154

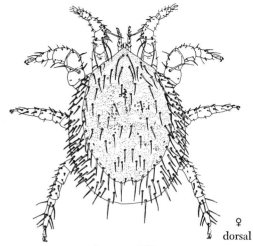

♀ dorsal

Figure 155

Figure 154 *Steptolaelaps liomydis* (Grant)

Idiosoma oval, dorsal plate widest at shoulders, dorsal setae stout and spine-like, tritosternum well developed with pair of lacinae narrowly elongate, pilose, divided almost to basal piece, holoventral plate of male entire including anal plate.

5b   Epigynial plate with only one pair of setae **6**

6a   Posterior setae of coxa II longer than the greatest diameter of the coxa. Large, coarse, hairy mites, mostly over 1 mm in length .... **Genus *Gigantolaelaps***

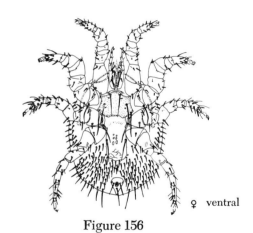

♀ ventral

Figure 156

Figures 155 and 156 *Gigantolaelaps mattogrossensis*

Dorsal plate with notched posterior margin, sternal plate 310-350 microns in length, coxa with both setae spine-like, sternal plate without small setae between anterior pair of setae.

Collected from *Oryzomys palustris palustris*, *Oryzomys palustris coloratus*, *Rattus norvegicus*, *Rattus rattus*, *Sigmodon hispidus komareki* in Florida, Georgia, South Carolina and Texas. The rice rat, *Oryzomys palustris* is considered to be the common host.

6b Posterior setae of coxa not longer than the greatest diameter of the coxa 7

7a Body oval or elliptical; legs I distinctly longer than legs II and III; posterior seta of coxa III similar to other coxal setae .... Genus *Haemolaelaps* Berlese

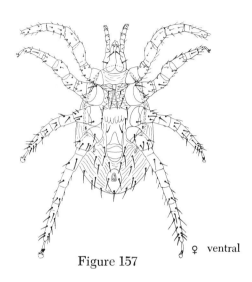

Figure 157 ♀ ventral

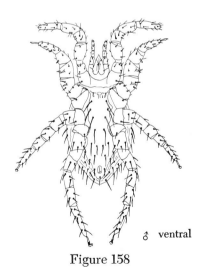

Figure 158 ♂ ventral

Figures 157 and 158 *Haemolaelaps geomys* Strandtmann

Female anal plate slightly arched, slender portion of the piled dentilis bent at right angles, legs IV longer than the dorsal plate, labium with a brush of long setae, pilies dentilis strongly inflated, body setae rather coarse. Male expansion of the labium with a brush of setae as in female, leg II with a heavy spine ventrally on each of the four apical segments. Nymphs distinguished by the presence of the labial brush.

Collected from *Cratogeomys castanops, Geomys bursarius, Geomys cumberlandicus, Geomys floridanus, Geomys lutescens, Geomys personatus, Geomys texensis, Geomys tuza, Thomomys bottae, Thomomys bulbivorus* in Florida, Georgia, Texas, Nebraska, Illinois, Oregon and California. It can be expected to be found wherever pocket gophers are present.

7b Body circular, female epigynial plate with one pair of setae, legs I subequal to legs II, posterior seta of coxa III thornlike, anal shield much broader, not pyriform .... Genus *Eubrachylaelaps*

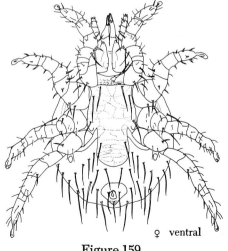

Figure 159 ♀ ventral

Figure 159 *Eubrachylaelaps circularis* (Ewing)

Female with sternal plate two-three and a half times as broad as long, without basal expansion of sternal plate setae, though slightly stouter than genitoventral setae, anal plate length subequal to width, genitoventral plate broadly rounded.

Collected from *Neotoma mexicana, Peromyscus boylii, Peromyscus californicus, Peromyscus hylocetes, Peromyscus maniculatus, Peromyscus oaxacensis, Peromyscus truci*, in Arizona, California, Colorado, and Utah.

Family Spinturnicidae

The mites of this family are exclusively parasitic on bats during all stages of their life cycle. They are normally associated with the wing membranes. They are blood feeders, have strong thick legs with heavy curved claws. The larvae are passed completely within the body of the female which gives birth directly to the protonymph.

Figure 160 *Periglischrus vargasi* Hoffman

Sternal setae of female on margins of shield, six pairs of dorsal opisthosomal setae in female, proximal dorsal setae of femur II, genu II and tibia II tiny, tritosternum absent, sternal shield longer than wide, epigynial plate long, slender.

Collected from *Leptongeteris navalis* in Texas.

1a Peritremes very long, extending from level of coxae IV to level of coxae I, two dorsal shields, tritosternum lacking .... **Genus** *Periglischrus*

1b Peritremes bent ventrad, single dorsal shield, tritosternum normally present .... **Genus** *Spinturnix*

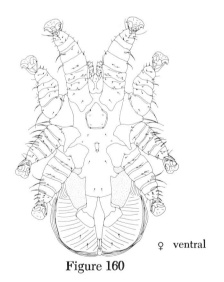

♀ ventral

Figure 160

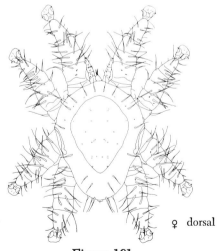

♀ dorsal

Figure 161

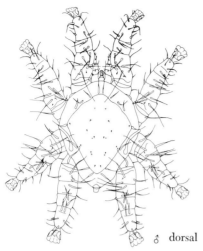

Figure 162

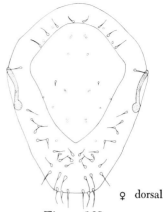

♀ dorsal

Figure 163

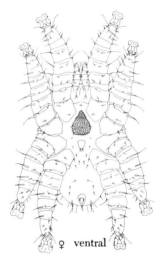

♀ ventral

Figure 164

Figures 161 and 162 *Spinturnix americanus* (Banks)

Ten to twelve dorsal opisthosomal setae in female, long posterolateral setae of tibia III and IV usually present, dorsal shield large, opisthosomal integument lightly striated, tritosternum small, epigynial shield longer than wide, narrow posteriorly.

Collected from *Myotis lucifugus, Myotis yumanensia, Myotis volaus, Myotis keenii, Myotis evotis, Myotis subcelatus* in Indiana, California, Alaska, Colorado, Illinois, Maine, Maryland, Massachusetts, Montana, Nevada, Oregon, Pennsylvania, Virginia, Wisconsin, Iowa, New York, Arizona, Arkansas, Delaware, Georgia, Kansas, New Jersey, Oklahoma, Texas, West Virginia.

Figures 163 and 164 *Spinturnix banksi* Rudnick

Thirty-two to thirty-six dorsal opisthosomal setae on female, about eight to ten dorsal opisthosomal setae on male, dorsal shield mended, epigynial shield small, pointed posteriorly, with mended anterior expansion, male anal shield ventroterminal, mended anteriorly, blunt posteriorly.

Collected from *Myotis grisescens* in Arkansas, Illinois, Missouri, Kansas, Oklahoma, and Tennessee.

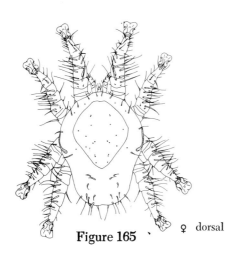

Figure 165　　♀ dorsal

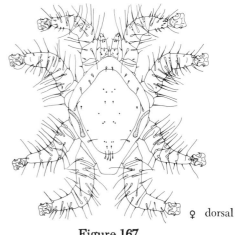

Figure 167　　♀ dorsal

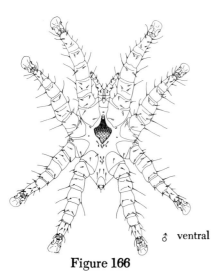

Figure 166　　♂ ventral

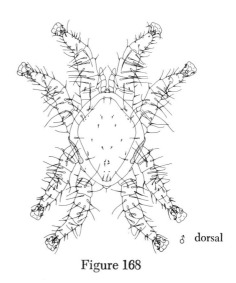

Figure 168　　♂ dorsal

Figures 165 and 166 *Spinturnix carloshoff-manni* Hoffman

Sixteen to twenty-five dorsal opisthosomal setae on female, lacking posterolateral setae of tibia III and IV. This species is very similar to S. *americanus* and the above structure will distinguish S. *carloshoffmanni* from S. *americanus.*

　　Collected from *Myotis velifer incautus* in Texas.

Figures 167 and 168 *Spinturnix bakeri* Rud-nick

Four dorsal opisthosomal setae in male and female, dorsal shield occupying most of po-dosomal surface, epignial shield small, deli-cately sclerotized, rounded posteriorly, male dorsal shield covering greater part of dorsal surface.

　　Collected from *Eptesicus fuscus bernar-dinus* in California, Arkansas, Minnesota, Pennsylvania, Arizona, Indiana, and Kansas.

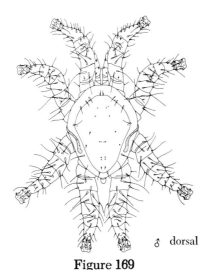

♂ dorsal

Figure 169

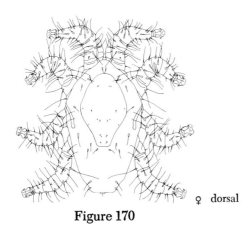

♀ dorsal

Figure 170

Figures 169 and 170 *Spinturnix orri* Rudnick

Eight dorsal opisthosomal setae on female, two dorsal opisthosomal setae on male, dorsal shield ovoid, epigynial shield small, delicately sclerotized with mended anterior expansion and pointed posterior apex, male dorsal shield covering greater part of idiosoma, single pair

of opisthosomal setae bordering posterior apex of dorsal shield.

Collected from *Antrozores pallidus pacificus, A. pallidus cantwelli, A. pallidus pallidus* in California, Oregon, Texas, Utah and Arizona.

1c   Single dorsal shield, tritosternum present, peritremes completely dorsal ....
.... **Genus *Paraspinturnix***

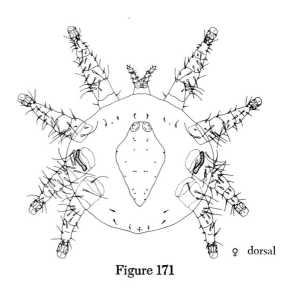

♀ dorsal

Figure 171

Figure 171   *Paraspinturnix globosus* (Rudnick)

Propodosomal integument with five pairs of setae, opisthosomal integument with two pairs of short setae and one pair of pores near posterior margin, dorsal shield widest at level of coxae III, epigynial shield narrow with wider anterior expansion, pair of short setae.

Collected from *Myotis sodalis, Myotis grisescens, Myotis velifer* in Tennessee, Indiana, Utah, Arizona, Oklahoma and Kansas.

Keys to the Common Hard and Soft Ticks of the Metastigmata

The families which include the creatures popularly known as "ticks" are very important to agriculture. Ticks are unattractive to the average person, and even more so to the agriculturalist; yet a knowledge of each species is essential if we are to continue to feed our ever growing population.

Wide distributions, great variation in size, tremendous reproductive capabilities, and resistance to control efforts are characteristics that unite to place them among the most formidable pests of domestic livestock. There is scarcely any domestic or wild animal that is not subject to their attack. Each tick, after having settled on its hosts, essentially turns itself into an automatic pump extracting the blood so vital to the life of the host. Injury to the host is aggravated in many cases due to the tick acting as a vector of certain disease organisms. Some species of ticks are restricted to a single host, others to a single genus or family of host animals. Many are extremely omnivorous and infest animals of unrelated groups. The ticks are generally distributed throughout the world wherever suitable host animals are found. Though many genera of ticks are native to certain regions of the world they have been distributed in commerce and are firmly established in other parts of the world.

Ticks are divided into two major groups, the "hard ticks" and the "soft ticks." The hard ticks contain two families, the Ixodidae and Nuttalliellidae. The soft ticks contain only the single family Argasidae.

During larval, nymphal, and adult stages most ticks are intermittent parasites, spending most of their existence on ground covered with small bushes and shrubs. Some species, however, such as the hard tick, *Boophilus annulatus*, pass most of their life on a single host. These are referred to as *one host ticks* and molt on the host. Favorable environmental conditions for ticks include the presence of suitable hosts for food, vegetation, and moisture. Certain species, however, may exist under environmental conditions where there is a scarcity of vegetation and moisture. Members of the family Argasidae are mainly nocturnal in habit, feeding rapidly during the night and concealing themselves during the day in crevices near the nest of its host.

Ticks are long lived. Some specimens have been kept under observation for seven years. The life cycle of *Dermacentor andersoni*, the Rocky Mountain spotted fever tick is approximately 20 months, however it may last up to three or more years depending on environmental stresses imposed.

Feeding may be accomplished by all three stages of the life cycle of a tick. Adults of the genus *Otobius* do not feed and whether nymph 1 ever feeds is not evident. The larvae and nymphs usually feed on small animals with the adults selecting larger animals. They attach to animals that come in contact with infested vegetation. Both sexes are blood suckers, and the adults usually require a feeding period of four to six days before copulation. The adult female and in some cases the nymph and larvae increase greatly in size because of the large amount of blood consumed. The male, as a rule, lacks the ability to extend its abdomen by taking in blood and remains small in size when compared with the engorged female. This is especially true of the hard ticks. The length of time required for engorgement of all stages and sexes depends on the particular species and stage of development. Some species are quick feeders requiring from 15 to 20 minutes to become fully engorged. Other species may require a week of continuous feeding to become fully engorged. Still others are intermittent feeders and will continue to attack the host over long periods of time. In most species engorgement of all stages and sexes is prolonged during low temperatures.

Upon completing her final blood meal, the engorged female tick drops off the host animal. She has the capacity to deposit from 2,000 to 18,000 eggs in masses on the ground, after a variable preoviposition period that is dependent on the temperature and species concerned. The soft ticks lay fewer eggs, 100 to 200, in several batches following successive blood meals. Tick eggs may be associated with the roots of grasses or among vegetable debris. In some species the egg is covered with a viscid secretion which prevents desiccation.

The life cycle of ticks vary both within and between families. The stages in the life cycle of ticks are egg, larva, nymph and adult. The larvae have three pairs of legs and no evidence of a tracheal system. They are active, usually attach to small animals, and

secure a blood meal, then drop off and molt to become a first instar nymph (in some species the larvae molt on the host and do not drop off). The nymphs have four pair of legs and a tracheal system but are without a genital pore. These in turn attach to an animal, engorge, drop off and molt to become adults. This complicated life cycle has given rise to the use of the phrases, "one-host ticks," "two-host ticks," and "three-host ticks" for certain species of ticks. The life cycle of a three-host tick begins when eggs hatch into active hexapod (six-legged) larvae. These larvae climb some nearby bush or shrub to await the passage of a suitable host. When a host brushes the plant the larvae move onto the host, find a suitable area on the body of the host for feeding, insert the mouthparts and feed for a period of time. The engorged larva drops off its host and molts on the ground giving rise to an eight-legged nymph. The nymph requires a blood meal and, like the larva, places itself on some type of vegetation to await the arrival of a suitable host. After the nymph has completed a blood meal it too drops to the ground and molts. From this molt the adult emerges and attacks yet another host. This completes the three-host cycle. Some tick species remain on the first host for the molt at the end of the larval stage, thus the same animal supports both larva and nymph. An additional host is required for the adult. This type of cycle is referred to as a two-host tick. The one-host tick completes its entire life cycle, larvae, nymph and adult on one host. The female is fertilized either before or after she becomes attached to the final host (usually after).

Various modifications of these cycles may occur, such as change of host, length of time on the host, number of molts and frequency of oviposition. For example, in the hard ticks there is but a single nymphal stage. The soft ticks may undergo several molts following successive blood meals.

To summarize, hard ticks of the genus *Boophilus* and some species of *Dermacentor* are one-host ticks. In two species of the genus *Rhipicephalus* and one species of *Hyalomma* both the larval and nymphal stages are passed on one host and adult stage on another host (termed two-host ticks). Members of the genus *Amblyomma* and many species of *Dermacentor* reflect the three-host type of life cycle. Small hosts moving in infected grassland areas pick up the larvae on all parts of their body; large animals have only their legs in contact with the grass thus providing a more limited site for host attachment. Therefore, host size appears to be significant; the larvae feed on small animals, nymphs parasitize larger animals and adults attack the largest animals. In the soft ticks, the female indulges in several blood meals and ovipositions as an intermittent feeder. In one genus of the soft ticks, *Otobius*, the adult does not feed; hence, it requires no blood meal preliminary to oviposition.

Knowledge of the specific habits have an important bearing on the role ticks play in the transmission of various tick-borne diseases. For example, a two- or three-host tick has the ability to pick up a pathogenic organism from one host and pass it along to a new host at a later stage in its life cycle. However, the one-host tick must transmit the infection through the egg stage to the next generation if it is to continue as a vector of that specific disease. This is referred to as transovarial transmission of a disease pathogen.

1a   Scutum present; capitulum terminal; mouthparts visible from dorsal aspect; palpi rigid, not fingerlike, only three segments clearly visible (terminal segment hidden in pit on third segment); pulvilli present (p. 103) ....
Family Ixodidae 5

1b   Scutum absent; capitulum ventral; mouthparts usually not visible from dorsal aspect; palpi flexed and fingerlike, with four visible subequal segments; pulvilli absent (p. 98) ....
Family Argasidae 2

2a   Body margin dorsally and ventrally separated by a definite sutural line. This line not obliterated when tick is fully gorged. Eyes always absent ....
.... Genus *Argas*

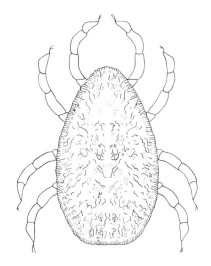

Figure 172

Figure 172   *Argas persicus* (Oken) Fowl tick

The body is usually oval, sometimes more elliptically shaped when fully mature. Postpalpal hairs present; hypostome apically notched and flattened margins having quadrangular plates.

Cosmopolitan species being restricted to warm, dry regions. United States—California across southwestern and southeastern states to Florida, Europe, Asia, Afric, Australia, Central and South America.

One of the important poultry parasites,

they are found in cracks of poultry houses, only the larvae may remain attached to the host for any long period of time. They are mainly night feeders, attacking a sleeping or nesting host. They attack man in some parts of the world producing serious symptoms that may require medical care. Economic losses as a result of this tick are considerable. Fowls are weakened through loss of blood and annoyance when attacked by large numbers the effects are sufficient to result in death. *A. persicus* transmits a spirochaetal disease to poultry. This flattened tick is well adapted to conceal itself in the woodwork of buildings where it lives gregariously.

2b **Body margin lacking sutural line separating the dorsal and ventral surfaces** 3

3a **Hypostome fully developed, either scoop shaped or otherwise formed, with teeth, nymphal stage without spinose integument** 4

3b **Hypostome vestigial in adults, integument granular, nymphal integument beset with spines, hypostome well developed** .... **Genus *Otobius* Banks**

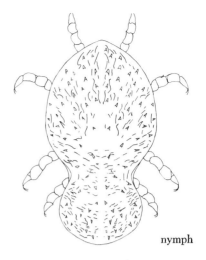

nymph

Figure 173

Figure 173 *Otobius megnini* (Duge's) Spinose Ear Tick

Adults and nymphs dissimilar; adults with integument granulated; nymphs striated and with spines. Hypostome of nymphs well developed; vestigial in adults without deuticles; larvae with striated integument, with bristle-like hairs, well developed hypostome and with six legs.

Distribution: Mexico, South America, South Africa, United States—common in parts of Texas, New Mexico, Arizona, and southern California, however, may occur where cattle and horses are found throughout the United States.

This is a two-host tick, the larva and nymph feed within the ears of their host. The adult is not parasitic nor does it take food. The common name is derived from the spines present on the nymphal instar. This stage and the larvae are the commonly collected forms. It is the nymphal stage in which this species is most easily distinguished from other ticks. The ear of the host can be thoroughly closed by an infestation of nymphs and larvae causing deafness in domesticated animals. This species is of exceptional interest due to its feeding habits. Other argasid ticks are intermittent feeders. *O. megnini* has only three molts, one larvae and a nymphal stage. Whether nymph one feeds is not known at the present time.

Otobius lagophilus Cooley and Kohls is similar to the well known spinose ear tick, *O. megnini*, but is separated by its smaller size, heavy "V" shaped spines missing and replaced by slender spines, denticles on hypostome 3/3 pattern instead of 4/4, legs more slender and spiracles of nymph mildly convex instead of conically protuberant as in *O. megnini*. The adults of this species are not parasitic and do not require any food. The nymphal stage has been found to feed on the face of its host rather than within the ear. They attack rabbits of the genus *Lepus*, the common jackrabbit is a common host. This species is recorded from Canada, and the United States—California, Colorado, Idaho, Montana, Nevada, Oregon and Wyoming.

4a Hypostome broad at the base and scooplike, body pointed at anterior end .... **Genus *Antricola***

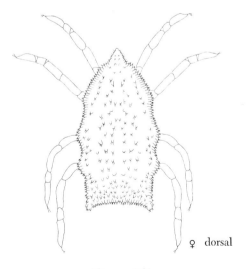

♀ dorsal

Figure 174

Figure 174 *Antricola coprophilus* (McIntosh)

The adult sexes are dissimilar, nymphs similar. Body shape pointed at anterior apex, two marginal projections above spiracles, tubercles large, moderate in number, of various shapes and sizes, those in median region almost a hemisphere, those on margins fused, directed outward at an angle, discs present, all coxae contiguous, cheeks absent, hypostome short, broad, rounded apically, convex ventrally and concave dorsally (scooplike).

Specimens of this species are rarely collected because of their small size and their habit of not clinging to their hosts. The mouthparts are adapted for quick feeding. They are found associated with bat caves, actual observations of them feeding on bats have not been made.

4b Hypostome variable but never scooplike, body thick, leathery, or flattened .... **Genus *Ornithodoros***

This genus contains a large number of species, and contains many of medical importance. In the western hemisphere five species are known to transmit disease, and in this connection it has been considered desirable to identify them by the following key from Arthur (1962).

1a Cheeks present, dorsal humps on legs absent 2

1b Cheeks absent, dorsal humps present or absent on legs 3

2a Disks large and conspicuous .... *O. talaje*

2b Disks small and inconspicuous .... *O. rudis*

3a Dorsal humps on tarsi I present 4

3b Dorsal humps on tarsi I absent ....
.... ***O. hermsi***

4a Eyes absent 5

4b Eyes present .... ***O. coriaceus***

5a Subapical dorsal protuberance on leg IV absent 6

5b Subapical dorsal protuberance on leg IV present .... ***O. nicollei***

6a Mamillae large, relatively few in number, not crowded .... ***O. turicata***

6b Mamillae small, many and somewhat crowded .... ***O. parkeri***

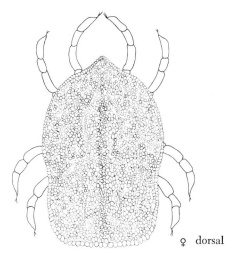

Figure 175

Figure 175   *Ornithodoros talaje* (Guerin-Méneville)

This species has a distributional range extending from California and Kansas to Argentina. In the United States it has been recorded from California, Arizona, Nevada, Kansas, Texas, and Florida. *O. talaje* is recorded from a wide variety of hosts. In the United States it is found associated on or with wild rodents. In Central America it is recorded as attacking man and found in the cracks of walls in human dwellings. Hosts include members of the rodent genera *Dipodomys, Citellus, Neotoma,* and *Mus. O. talaje* is a vector of relapsing fever in the United States.

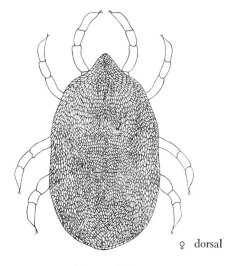

♀ dorsal

Figure 176

Figure 176   *Ornithodoros hermsi* Wheeler, Herms and Meyer

This tick was demonstrated by laboratory experiments on a monkey to be a vector of relapsing fever. It is restricted to the United States in its distribution, having been collected from California, Colorado, Idaho, Nevada and Oregon. Host records include only cases where it has been found to attack man.

All collections are from bedding nests of chipmunks, hollow logs and cabins. No recent cases of relapsing fever traceable to O. *hermsi* have been recorded.

Ornithodoros concanensis Cooley and Kohls

Host and distribution of this tick is restircted to Arizona and Texas from bat caves. It is recorded as quite similar to O. *talaje* and O. *Kelleyi* by the original describers. Ticks were collected on guano and rock crevices and not from bats themselves.

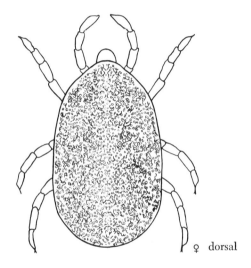

Figure 177

Figure 177 *Ornithodoros parkeri*

This species is widely distributed throughout the western half of the United States—California, Colorado, Idaho, Montana, Nevada, Oregon, Utah, Washington, and Wyoming. Host records include species of *Citellus, Lepus, Cynomys,* all rodents. The burrowing owl, *Speotyto cunicularia* is considered to be a natural host of O. *parkeri.*

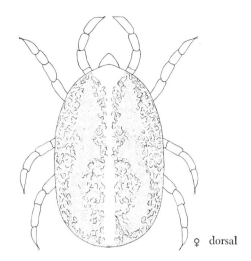

Figure 178

Figure 178 *Ornithodoros turicata* (Dugés)

Ranging throughout the south central United States, this tick includes a wide variety of hosts, including reptiles, birds, and mammals. O. *turicata* has been collected from rattle snakes, woodrats, rabbits, pigs, cattle, horses, man, burrowing owls, and terrapins. It is reported to be the cause of plague among pigs. The bite of this tick is extremely painful and often followed by serious secondary consequences, leading to the formation of gangrene on the skin in pigs. This species is the only known vector of human relapsing fever of man in portions of Kansas, Oklahoma, Texas, and possibly other areas in the southwestern United States. It has been recorded from Utah, Colorado, Kansas, Oklahoma, Texas, New Mexico, Arizona, California and Florida.

In addition to the 7 *Ornithodoros* species cited in the key, 12 other species have been recorded from the United States, Central America and Cuba. Most of these are parasites of members of the order *Chiroptera* (bats), and have been collected from bat guano.

5a **Anal groove distinct extending and surrounding the anus in front** .... **Genus *Ixodes***

Anal grooves surrounding the anus anteriorly and usually uniting in an arch. Sexual dimorphism pronounced. Inornate, without eyes and without festoons. Palpi and basis capituli variable in form. Articulation between palpal articles I and II movable. Coxae either with or without spurs; spurs on coxa I when present are variable. Male with ventral plates present—one median, one anal, two adanals; sometimes two epimeral plates; pregenital plate sometimes present.

The members of this genus will prove to be difficult for the beginner. The work of R. A. Cooley and Glen M. Kohls, *The Genus Ixodes in North America,* is an indispensable publication when members of this genus are encountered from North America.

5b **Anal groove distinct or indistinct, never extending anteriorly around anus, contouring the anus behind if distinct** **6**

6a **Without eyes** **7**

6b **With eyes (except certain species that may have eyes distinct, absent or vestigial in same species population)** **8**

7a **Article (segment) projects or acutely produced laterally beyond the basis capituli, about twice as broad as long** .... **Genus *Haemaphysalis***

Small ticks; some females are fairly large when engorged. Scutum ornate; without eyes; female lacking lateral grooves. Festoons usually eleven in number. Trochanter I with a bladelike, dorsal, retrograde process. Coxa I never bifid. Palpi usually short, conical, and wide, projecting laterally beyond basis capituli which is rectangular dorsally. Sexual dimorphism slight; ventral plates and shields on male absent. Spiracular plates of male oval or comma-shaped; in female rounded or oval.

This genus is represented in the United States by *Haemaphysalis leporispalustris* (Packard), the rabbit haemaphysalid.

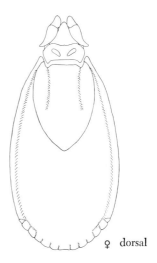

♀ dorsal

Figure 179

Figure 179   *Haemaphysalis leporispalustris* (Packard) Palpi short, conical, article two projecting laterally beyond basis capituli.

Hypostome dentition 3/3; scutum inornate; cervical grooves of male long, convergent, deep. Palpi of both sexes longer than broad; article two projecting laterally, slightly recurved. Coxa I with short external and internal spurs; coxae II-IV with small spurs. Trochantal spurs slightly developed. Spiracle large with pointed dorsal process.

This genus is most frequently found in the United States attached to rabbits. It is widely distributed throughout North and Central America. It has been collected from species of birds and occasionally domestic mammals such as the horse, cat and dog. It rarely feeds on humans but is sometimes found in houses.

7b **Article two without projection of basis capituli, about twice as long as broad** .... **Genus *Apanomma***

Small, eyeless; article two without projection of basis capituli, about twice as long as broad; ornate or inornate. Males without adanal shields. Spiracular plates sub-triangular or comma-shaped.

Members of this genus are found associated with large snakes and lizards. They are host-specific, rarely feeding on animals other than their normal host. This group was originally treated as belonging to the genus *Amblyomma*.

8a **Palpi as broad as or broader than their length** **9**

8b **Palpi longer than broad** **12**

9a **Basis capituli rectangular dorsally, usually ornate, festoons present; common in United States in upper Midwest** .... **Genus *Dermacentor***

Ornate, with eyes and festoons (11 in number). Basis capituli quadrangular dorsally. Hypostome with three rows of denticles on each side of the median line. Palpi short, broad or moderate in width, and with postero-dorsal elevation present; article one fused with article two. Coxae I to IV increasing in size progressively with coxa IV very large; coxa I bifid. Spiracles suboval or comma-shaped. Post anal groove present (sometimes indistinct). Male with no ventral plates or shields.

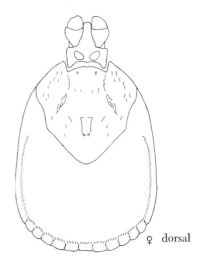

♀ dorsal

Figure 180

Figure 180 *Dermacentor andersoni* Stiles Scuta with very deep, large punctations, spiracular plate subcircular, with spiracular opening sub-central.

Hypostome 3/3, eyes present, cornua broad, rounded apically, broader basally than long. Scutum with punctations of two sizes, small punctations which cover the entire scutum and larger punctations which are marginal and sometimes arranged in diagonal rows. Spurs on coxa I well developed, proximal edges either parallel or only slightly divergent. External spurs present on coxae II, III, and IV, moderate in length; internal spurs on coxae II and III short, broad, absent on coxa IV. Male distinguished by spiracular plate with dorsal prolongation not long and narrow, and with larger punctations in scutum very large and deep, as in female.

The cornua, deep punctations and size of goblets serve to differentiate this very important economic species. The short or moderate cornua distinguishes *D. andersoni*, *D. variabilis* and *D. hunteri* from the other North American species. Both *D. andersoni* and *D. hunteri* have spiracular goblets which are moderate in size and number. The deep punctations on the margin of the scutum distinguishes *D. andersoni* from *D. hunteri*.

This is one of the most familiar ticks in North America due to its association with disease pathogens. *D. andersoni* is called the Rocky Mountain spotted fever tick. It is distributed generally in the north western sections of the United States. The adults of this species prefer large animals such as cattle, horses, sheep, dogs, big game, other large wild animals and man. The recorded host list is long and probably any large mammal is susceptible to its attack. The larvae and nymphs also have a wide range of host species. Like the adults it is believed that any mammal may serve as a host for either the larvae or nymph.

D. andersoni serves as a vector for Rocky Mountain spotted fever, "Q" fever, tick paralysis, tularemia, and Colorado tick fever. It also is an annoyance to any host due to its blood sucking habit. It is the main vector of Rocky Mountain spotted fever, *Rickettsia (Dermacentroxenus) rickettsii.* The pathogen is maintained in *D. andersoni* from generation to generation by transmission of the microorganisms through the eggs. The disease has a high mortality rate in man among unvaccinated and untreated cases, ranging from 65-70 percent.

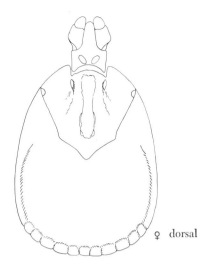

♀ dorsal

Figure 181

Figure 181 *Dermacentor variabilis* Say Spiracular plate with goblets very numerous and small, oval, with dorsal prolongation.

Hypostome 3/3, eyes present, variably ornate on scutum, cornua shorter than their basal width, rounded apically. External spurs present on legs I-IV, very large triangular internal spur on coxa I. Spiracular plate with blunt dorsal process, goblets very small and numerous. These small goblets distinguish the species from all other members of the genus *Dermacentor* from the United States.

Dermacentor variabilis is well known and called the American dog tick. It is widely distributed east of the Rocky Mountains, in parts of California, Mexico, and Canada. In some of these regions it replaces *D. andersoni*.

This species is the principal vector of Rocky Mountain spotted fever in the central and eastern sections of the United States. It can transmit tularemia, bovine anaplasmosis and cause canine paralysis. The encephalomyelitis has been transmitted under laboratory conditions and has been implicated as a possible cause of tick paralysis in the eastern section of the United States. It has been demonstrated that certain parasites can be passed from mother to young by transovarial transmission.

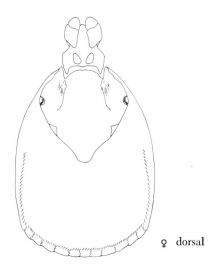

♀ dorsal

Figure 182

Figure 182 *Dermacentor occidentalis* Marx Cornua long, spiracular plates oval, with dorsal prolongations.

Hypostome 3/3, eyes present, very ornate on scutum. Cornua as long as their basal breadth, narrowing at apex. Legs very ornate. Internal spurs on coxa III of female reduced to marginal saliences. Coxa I with subequal internal and external spurs. External spurs of coxae II-IV longer than their basal breadths, tapering to apex.

The long cornua of *D. occidentalis* is a positive means of separating it from *D. variabilis*, *D. andersoni* and *D. hunteri* of North American species. This species is also separable from *D. parumapertus* and *D. halli* in having the spurs on coxa I with proximal margins parallel or only slightly divergent. From *D. albipictus* it may be distinguished by the oval spiracular with dorsal prolongations. This is the well-known western dog tick or Pacific tick. It is common in California, Oregon, ranging throughout the coastal region of the west. Having first been collected on deer which at times becomes heavily infested. It is a major pest of cattle, horses, man, dogs, sheep and rabbits. Nymphs and larvae are found on many species of smaller animals such as field mice, skunks, rabbits and ground squirrels.

D. occidentalis is a possible natural carrier of spotted fever virus, has been incriminated as able to transmit tularemia, bovine anaplasmosis, and "Q" fever in the western United States.

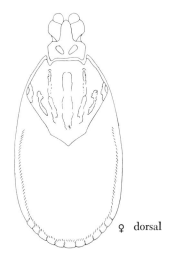

♀ dorsal

Figure 183

Figure 183 *Dermacentor albipictus* Packard Spiracular plate oval, without dorsal prolongation and with goblets few and large.

Hypostome 3/3, eyes present, cornua well developed, apically rounded. Cervical grooves shallow, varying in length in different specimens. Coxal spurs on coxa I bifid, slightly divergent at their apices, thin inner margins parallel. External spurs on coxae II, III, and IV well developed, twice as long as broad, narrowing apically. Internal spurs on coxae II and III pointed. Spiracular plate oval, without dorsal prolongation, goblets few and large. Male internal spurs present on coxae II and III. *Dermacentor albipictus,* the winter horse tick, is a one-host tick. It has been recorded from New York, Pennsylvania, Wisconsin, Tennessee, Oklahoma and Texas. The hosts include moose which is the type host, deer, *Cervus virginianus,* elk, *Cervus canadensis,* antelopes, *Antilocapra americana* and the Rocky Mountain sheep, *Ovis canadensis.* The hosts most often infested with *D. albipictus* are horses and cattle. The larvae or nymph remain on their host throughout the winter with the adults emerging in the early spring.

It is believed that *D. albipictus* is a vector of various pathogens of wild game, especially large game animals, causing severe loss of deer, etc.

9b Basis capituli hexagonal dorsally, inornate **10**

10a Anal grooves distinct, with eyes, festoons, coxa I bifid, male with pair of adanal shields and usually with a pair of accessory adanal shields; spiracular plates bluntly or elongately comma-shaped .... **Genus *Rhipicephalus***

Inornate; with eyes and festoons. Palpi short, broad; posterolateral angles of segment two do not project strongly. Basis capituli hexagonal dorsally; cornua distinct. Coxa I bifid; external spur straight. Males with a pair of adanal shields and usually a pair of accessory shields. Spiracular plates comma-shaped.

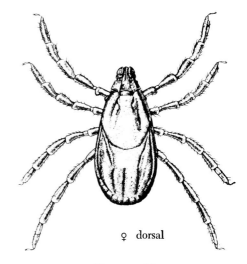

♀ dorsal

Figure 184

Figure 184 *Rhipicephalus sanguineus* (Latreille) External spur of coxa I shorter than internal spur, infra-internal setae of palpal segments one and two massive, split at the end or feathered. Male adanal shields narrow, being about three times as long as broad.

This is the well known brown dog tick that has been widely distributed throughout the world. It is often called the kennel tick in many parts of the world. It is a native of Africa and has followed its primary host, the domestic dog, to all parts of the world.

R. sanguineus is the only member of the genus *Rhipicephalus* that occurs in the United States. Dogs are its principal host, however it has a wide range of hosts and infrequently attacks man when in close association with dogs and their bedding places. In the United States it is found mostly in the south, south central, and southwestern portion of the country.

10b Anal grooves faint or weakly indicated in males, absent in females, eyes present, festoons absent, lightly chitinized. Spiracular plates round or oval **11**

11a Light colored, palpi ridged dorsally and laterally; male with contiguous leg segments, integument and legs without conspicuous long hairs .... **Genus *Boophilus***

Poorly sclerotized. Scutum inornate; with eyes; scutum in females quite small. Palpi very short, compressed, with dorsal and lateral transverse ridges on the articles. Basis capituli hexagonal dorsally. Anal groove in male faintly indicated; obsolete in the female. Males with adanal shields. Spiracular plates rounded or oval in both sexes. Festoons absent.

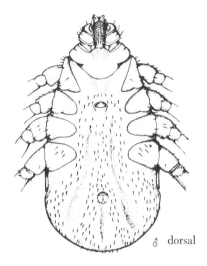

Figure 185

Figure 185 *Boophilus annulatus* (Say) Male without caudal appendage, female coxa I with shallow, rounded emargination separating internal and external spur.

Hypostome broad; dentition 4/4. Eyes present. Basis capituli very broad, posterolateral angles with slight protuberances. Palpi short, broad; article two broader than long, with asymmetrically convex apex. Female scutum longer than broad, cervical grooves shallow. Coxa I of female triangular, with external and internal spurs broadly rounded, equal in size, separated by shallow rounded emargination. Spiracular plate elongate oval; goblets small and numerous. Male without caudal appendage.

This is the famed Texas cattle fever tick and is normally restricted to North America south of the 37° latitude into parts of Mexico. It presumably began as a parasite of the deer and buffalo and subsequently transferred to domestic cattle. It is the vector of a protozoal infection called Texas or red water fever. The causal organism belongs to the family Babesidae.

11b Dark colored, palpi not ridged dorsally and laterally, male with noncontiguous leg segments, integument hairy and legs with conspicuous long hairs .... **Genus *Margaropus***

This genus was regarded as occurring in South America, however, this is now considered to be erroneous. Though related to the genus *Boophilus*, members of this genus are only found in Africa and Madagascar. The genus consists of only two species, *M. winthemi* Karsch of Southern Africa and Madagascar and *M. reidi* Hoogstraal from the Sudan.

12a Scutum ornate (ornamentation may be faded in poorly preserved specimens), males without adanal or subanal shields. Both sexes palpal segment two at least twice as long as segment three in species collected in United States. Festoons regular not coalesced .... **Genus *Amblyomma***

Usually ornate with dark spots and stripes on a pale background. Eyes and festoons present. Palpi usually long with article two especially long. Articulation between palpal articles one and two movable (not fused). Basis capituli variable in shape, often sub-quadrangular or sub-triangular. Males without the adanal shields, ventral plaques, scutes may be present. Spiracular plates sub-triangular or comma-shaped. Anal groove embracing the anus posteriorly.

The genus is large in numbers of species, there being approximately 100 definitely established species. Its geographical distribution is worldwide, however, about half of the known species are found in North and South America. Also the host range is large including not only mammals, but reptiles, and birds (nymphs and larvae).

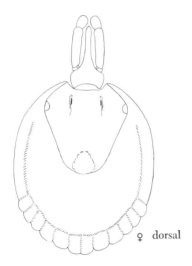

♀ dorsal

Figure 186

Figure 186 *Amblyomma americanum* (Linnaeus) Scutum with an ornate circular spot near the posterior margin.

Hypostome dentition 3/3, scutum with ornate spot at posterior margin. Scutum with cervical grooves short and deep, punctations numer-ous. Coxae I with two spurs, external spur long, pointed. Coxae II and III with a single broad platelike spur, coxa IV with a single triangular, platelike spur.

The distribution of *Amblyomma americanum* is predominantly southern. There are records from Michigan to New York; however, if present, it is extremely scarce in northern sections of the United States. It is common in Texas and should be expected to be found in Arkansas, Georgia, Louisiana, Mississippi, Missouri, Oklahoma, and Tennessee. This tick shows a wide range of hosts which include birds.

Amblyomma americanum is commonly called the lone star tick because of a single spot of ornate coloring on the scutum. *A. americanum* is known to be a vector of American spotted fever (Rocky Mountain spotted fever) and replaces *D. andersoni* as the main vector of this disease in the southern United States. This tick is also a possible cause of tick paralysis and a vector of "Q" fever.

12b   Scutum inornate, males with adanal and subanal shields. Palpal segment two less than twice as long as segment three. Festoons irregular, particularly coalesced .... Genus *Hyalomma*

Heavily sclerotized forms. Ornate or inornate, sometimes ornamentation confined to legs. Eyes present; without festoons. Palpi mostly long and thick. Articulation between palpal articles one and two movable. Basis capituli sub-triangular or sub-quadrate dorsally.

A very complex genus restricted to regions other than the United States.

Keys to the Common Neck Stigmatids of the Prostigmata

1a Body wormlike, elongated and annulated 2

1b Body not wormlike, rounded 4

2a Adult with only two pairs of legs; a transverse genital aperture placed behind second pair of legs; all plant feeders .... **Eriophyidae**

The species of this family are very small mites with a reduction of the number of legs to four. The recent work of Jeppson, L.R., H. H. Keifer and E. W. Baker, *Mites Injurious to Economic Plants*, should be used in identifying members of this family. They are all plant feeders and some species are gall makers.

2b Adult with 4 pairs of legs, these may be much reduced but present 3

3a Female genital aperture between legs IV; genital discs absent; body and leg setation reduced or absent; leg segments telescoped; parasites of the skin of animals (p. 192) .... **Demodicidae**

3b Female genital aperture behind legs IV; three genital discs; free-living .... **Nematalycidae**

This is a unique family in regard to its morphological structures. The two species recorded within the family are wormlike, elongate forms and have been collected from pasture soils in the United States. Members of the family Nematalycidae are soft bodied, with the female genital aperture behind legs IV; three pairs of genital discs; four well-developed pairs of legs; body annulate, elongate, wormlike. There is only one genus that is characterized by the structures of the family .... Genus *Nematalycus*.

♀ lateral

Figure 187

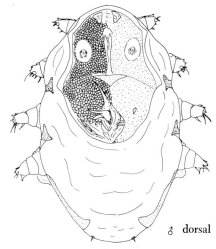

♂ dorsal

Figure 188

Figure 187 *Nematalycus strenzkei* Cunliffe

Chelae tiny, opposed; sicklelike claws and empodium on tarsus I; claws absent on legs II, III, IV, with broad, distally hooked, rayed empodia; genital opening with pairs of short simple setae; three pairs of genital suckers.

Collected from Oklahoma, Colorado, Texas and South Dakota from pasture soil.

4a   Palpi one-segmented and fanglike; gnathosoma tubular, without discernible chelicerae; found in cloaca of aquatic turtles .... **Cloacaridae**

This family contains a group of mites that are highly modified obligate parasites of turtles. They are found in the cloaca of their host. They are minute, soft-bodied, oval mites with greatly reduced legs and mouthparts. They have bright red eyes and a well developed dorsal shield with a complex reticular pattern. There is only a single genus contained in the family .... Genus *Cloacarus*.

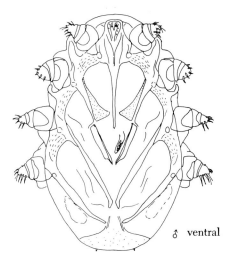

♂ ventral

Figure 189

Figures 188 and 189 *Cloacarus faini* Camin and Singer

Gnathosomal apodemes join those of legs I which fuse medially behind legs I, extending posteriorly to posterior margins of legs III; coxal apodemes II remain separate; male with reversible, bifid aedeagus between legs III.

Collected in Kansas from a female snapping turtle, *Chelyda serpentina.*

4b Palpi not fanglike, gnathosoma not tubular **5**

5a Females with from zero to three pairs of legs; males with three to four pairs of legs; pseudostigmata absent; parasitic on insects .... **Podapolipidae**

The family Podapolipidae contains highly evolved and specialized species that parasitize insects. The family is distinguished by the reduction of the number of legs in the adults. Baker and Wharton (1952) referred to this family as showing the greatest degeneration or specialization of all the Acarina. At the present time work is underway in an attempt to establish generic limits within the family. Only a single species is here illustrated to show the degeneration of the female, male, and larviform stages.

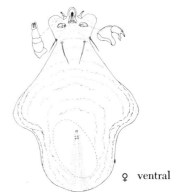

♀ ventral

Figure 190B

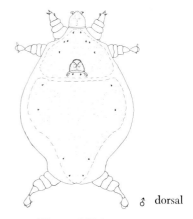

♂ dorsal

Figure 191A

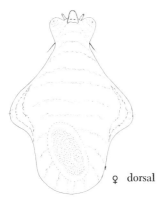

♀ dorsal

Figure 190A

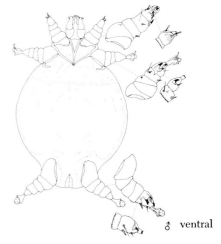

♂ ventral

Figure 191B

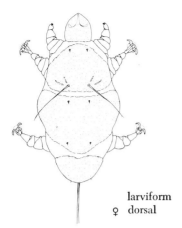

larviform
♀ dorsal

Figure 192A

larviform
♀ ventral

Figure 192 B

Figures 190, 191, and 192 *Coccipolipus hippodamiae* (McDaniel and Morrill)

Male with coxa III pore separated from coxa III seta by a distance nearly equal to the distance between coxae III; larviform female; opisthosomal setae present, long; three pairs of legs; claws on legs II and III with one tine, well sclerotized; coxa II setae further removed from midline than coxae I setae by about diameter of circle at base of setae; adult female with opisthosomal setae absent; one to two pairs of legs, sucking disc on leg I absent; tectum tonguelike; two spines on tarsi II; idiosoma with an interior and posterior pair of bulges.

5b   Females and males normal four pairs of legs; pseudostigmata present or absent **6**

6a   Dorsal body plate rooflike over gnathosomal region; leg IV ending in several whiplike setae; tarsus elongated, attenuated; with or without claws; associated with insects (p. 203) .... Scutacaridae

6b   Dorsal body plate not rooflike over gnathosomal region; leg IV not attenuate, without whiplike setae **7**

7a   Empodium membranous flaplike on legs II; claws arise from membranous base or absent with suckerlike empodium; pseudostigmatal organ present, fanlike on adults; palpi lying closely appressed to gnathosoma **8**

7b   Without above combination of characters **9**

8a   Legs IV of female without claws or empodia; with two terminal whiplike setae; male with legs IV modified with a single claw (p. 134) .... Tarsonemidae

8b   Legs IV of female with claws and empodia or claws absent with sucker-like empodium; idiosoma elongate or oval, greatly enlarged in gravid individuals; small mites associated with insects or plants (p. 142) .... Pyemotidae

9a   Palpal thumb-claw process distinct **10**

9b Palpal thumb-claw process indistinct or absent 30

10a Setae of body giving a hairy appearance, may be quite long or short, but densely covering the body; larvae parasitic, adults and nymphs free-living predators 26

10b Setae of body not giving the mite a hairy appearance, number being relatively few when compared with those mites having a hairy appearance; larvae, nymphs and adults live in similar habitat 11

11a Dorsal shields several in number, strongly armored; setae on leg I arranged on innersurface in a rakelike fashion; chelicerae short, thick, with strong sickle-shaped movable chelae and weak, fixed chelae; free-living (p. 188) .... Caeculidae

11b Dorsal shields if more than one not strongly armored; setae of leg I not rakelike; chelicerae without strong sickle-shaped movable chelae and weak fixed chelae 12

12a Chelicerae are hinged at base, able to move scissorlike laterally over gnathosoma 13

12b Chelicerae not hinged at base, not capable of movement in a scissor-like movement over gnathosoma 16

13a Palpal thumb long and well-developed with a prominent thornlike spine at apex; chelicerae with a distal hook; free-living .... Anystidae

Members of this family are long-legged, normally red in color, fast moving and predaceous on other small mites and insects. Because of their so-called whirlagig running pattern, they are difficult to capture. Relatively few workers have worked in this family in the United States. Cunliffe (1957) described a new genus and species and placed it in a new subfamily, the Adamystinae. It was collected from lodgepole pine cones in California. Banks (1894) described a species from New York. Baker and Wharton (1952) reported that a Venezuelan species of Anystidae had been collected in Florida. Krantz (1970) reported that a species of the genus *Tarsotomus* was a regular inhabitant of farm storage bins in Oregon. Cunliffe (1957) stated much remains to be done within this family at the generic and specific levels, and that descriptions are vague and synonyms appear to be inevitable.

13b Palpal thumb shorter, not long and prominent; chelicerae without a distal hook at apex 14

14a Claws of legs I-II serrate (rayed); three pairs of weakly developed genital discs .... Teneriffiidae

Members of the family Teneriffiidae are medium sized mites. The outstanding character of the family is the broadly bipectinate claws of the first tarsus. McDaniel et al. (1976) recognized two genera for the family, establishing a habitat preference as well as structural differences between the two genera. Only the genus *Parateneriffia* and the species *P. uta* (Tibbetts) have been recorded from the United States. The members of the genus *Parateneriffia* are terrestrial in habit in contrast to members of the genus *Teneriffia* which are semi-marine. The species *P. uta* (Tibbetts) is widely distributed in the United States from

Arizona to South Dakota in association with western range pastures .... Genus *Parateneriffia*.

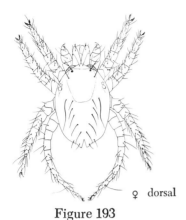

♀ dorsal

Figure 193

Figure 193 *Parateneriffia uta* (Tibbetts)

Coxa with four or less setae; palp genu with a process; dorsal median setae do not overlap; legs not longer than body; integument finely striated except for podosomal shield; two pairs of widely separated eyes; podosomal shield with one pair of pseudostigmatic setae; palpi five segmented; femur possesses a plumose setae on inner dorsal surface; genu bears a nude dorsal setae; a thumb-claw process on inner dorsal surface; hypostome with two pairs of short blunt papillae; genital opening with discs, genitalia internal, covered by elongated flaps associated with setae.

Collected from Utah, Texas, Arizona and South Dakota.

14b Claws of legs I-II not serrate 15

15a Tarsal claws present, with tenent hairs, without empodia; parasites of lizards and arthropods (p. 195) .... Pterygosomidae

15b Tarsal claws may be absent; if present, without tenent hairs; empodia present or absent; peritremes chambered; free-living .... Pseudocheylidae

Members of this family are small rhombiform or elongate mites with chambered peritremes which may be cervical or free. They are often associated with trees or moss and are only rarely collected. *Pseudocheylus* is the only genus known from the United States. .... Genus *Pseudocheylus*

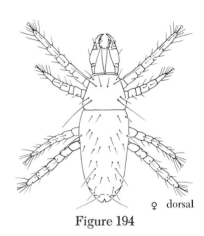

♀ dorsal

Figure 194

Figure 194 *Pseudocheylus sp.*

Peritremes free at distal ends or at antero-lateral projections; palpal tarsus reduced to a plate; claws and/or empodia on annulated stalks. According to Baker and Atyeo (1964) the absence of claws on pretarsus I is sufficient to distinguish *P. americanus* from the only other species in the genus, *P. biscalatus* Berlese from Brazil.

Collected in Illinois under the bark of a hard maple tree.

16a Legs III and IV without claws, ending in long, whiplike setae, legs I and II with claws and rayed empodia; parasites on birds .... Harpyrhynchidae

16b Legs III and IV not ending in long whiplike setae in place of claws 17

17a Pedipalpal free segments broadly articulated with basis capitulum, palp tarsus short, knoblike, sensilla conspicuous consisting of two large pectinate or comblike setae (combs); two smooth sicklelike setae (sickles), and a small inflated peglike setae (solenidion); free-living predators or associated with birds and mammals (p. 153) .... Cheyletidae

17b Without above combination of characters 18

18a Chelicerae long, recurved, whiplike, genital opening transverse; plant feeders 19

18b Chelicera not long, recurved and whiplike; genital opening longitudinal 22

19a Ocelli present on propodosoma 20

19b Ocelli absent; elongate striate forms with widely separated coxae II-III .... Linotetranidae

Linotetranids are small mites lacking eyes which distinguishes them from other closely related families. The chaetotaxy of the dorsum is also distinctive in that the propodosoma bears four pairs of setae, and the hysterosoma bears 17 or 18 pairs of setae. The genital plate of the female is triangular, longitudinally divided, and provided with three pairs of setae. The phallic organs of the male consist of a pair of strongly recurved genital stylets and a slender sclerotized aedeagus, there are six pairs of genito-anal setae. .... Genus *Linotetranus*.

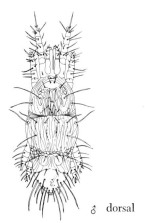

♂ dorsal

Figure 195

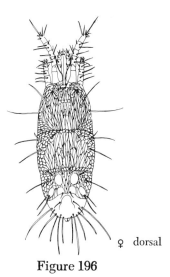

♀ dorsal

Figure 196

Figures 195 and 196   *Linotetranus achrous* Baker and Pritchard

Caudal division of dorsum of opistosoma with seven pairs of setae, anterolateral pair lacking, propodosoma with fine longitudinal striae, four pairs of setiform dorsal propodosomals, opisthosomal dorsum with transverse row of six setae along anterior margin, female tarsi III and IV without sensory pegs, genital plate triangular, with a pair of mediolateral setae on venter and three pairs of anal setae.

Collected from California on *Distichlis spicata*.

20a Claw terminating in a pair of tenent hairs or bordered with combs of tenent hairs; caudal aspect of idiosoma without highly modified setae; plant feeder (p. 141) .... Tetranychidae

20b Claw with a series of comblike tenent hairs, empodia similarly ornamented; caudal aspect of idiosoma with a series of flagelliform or dendritic setae 21

21a Dorsum of hysterosoma with 36 fan-shaped setae, caudal aspect of isiosoma with a series of long flagelliform setae; plant feeders .... Tuckerellidae

This group of mites has been classified within the family Tetranychidae. Like the tetranychoids, they are plant feeders. They are readily distinguished from other tetranychoid mites by the dorsal chaetotaxy of the body. There are normally fanlike dorsal propodosomals and 36 pairs of fanlike dorsal hysterosomals. There are five or six pairs of flagellate setae caudally, the palpus is distinctive in that the fifth segment is long and slender and bears only two sensory rods and three tactile setae. .... Genus *Tuckerella*

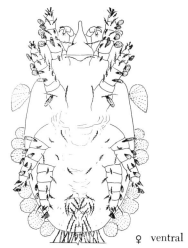

♀ ventral

Figure 197

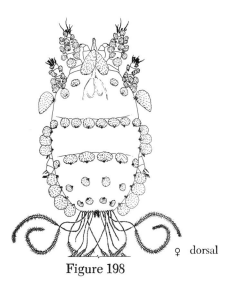

♀ dorsal

Figure 198

Figures 197 and 198 *Tuckerella hypoterra* McDaniel and Morihara

21b Dorsal setae all simple; caudal aspect of idiosoma with a series of short dendritic setae; claws each with comblike tenent hairs; plant feeders .... Allochaetophoridae

This family contains only the single genus and species, *Allochaetophora californica* Mc-Gregor. The species was first described within the family Tetranychidae .... Genus *Allochaetophora*

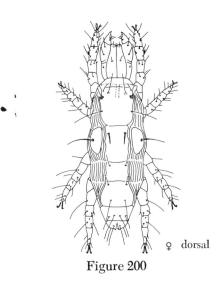

♀ lateral

Figure 199

Figure 199 *Allochaetophora californica* McGregor

Collected from California on Bermuda grass.

22a **Hysterosomal shield divided into four dorsomedial shields; propodosoma with a pair of clavate pseudostigmatic organs, with no more than two solenidia on tarsi I; without genital discs** .... **Tarsocheylidae**

The members of this family were formerly assigned to the family Pseudocheylidae. They are elongated forms with dorsal idiosomal shields, with the palpal tarsus reduced; no femoral division on the legs; genital discs are absent; and with small simple peritremes. They are associated with humus, rotten detritus and moss. Only the genus *Hoplocheylus*

Atyeo and Baker has been recorded from the United States. .... Genus *Hoplocheylus*

♀ dorsal

Figure 200

Figure 200 *Hoplocheylus discalis* Atyeo and Baker

Palpal tarsus indistinguishable; leg I without empodia; chelicerae medially fused or with form of stylophore; five mid-dorsal plates. Tarsus I with distal solenidion not extending beyond apex of tarsus; palpal tibia with three spines, two of which are blunt and subequal, the other platelike; tibiotarsus with five simple setae; one podiform setae; one small solenidion; and three subterminal spines.

Collected from Nebraska, Kansas from tree hole and rotting log.

22b **Hysterosomal shield not divided as described above; empodia present with tenent hairs** **23**

23a **Coxae II-III in contact; female genital and anal openings slightly separated; free-living** .... **Raphignathidae**

The family Raphignathidae contained many genera that have recently been placed in new families established for these genera. The genus *Raphignathus* and its species now comprise a closely allied group that defines the family Raphignathidae. They are small mites with three to four podosomal plates and one large opisthosomal plate. There may be two small plates at the posterior end of the podosoma. The eyes are in the lateral podosomal plates. Outstanding characters of this family are the presence of cervical peritremes and the contiguous coxae.

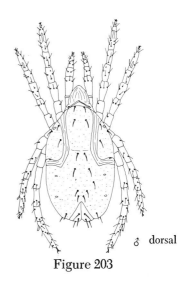

♂ dorsal

Figure 203

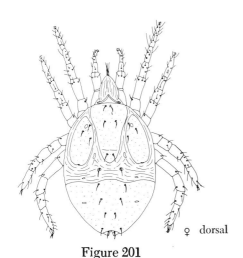

♀ dorsal

Figure 201

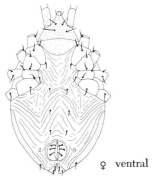

♀ ventral

Figure 202

Figures 201, 202, and 203 *Raphignathus collegiatus* Ateyo, Baker and Crossley

Dorsum of propodosoma with one medial and two lateral plates separated by finely striated areas; one large plate on opisthosoma which curves to venter, one plate between coxae I, one pair of external genital plates each with three setae, all plates heavily punctate.

Collected from Maryland under bark of horse chestnut tree and from Illinois and South Dakota.

23b   Coxae I-II distinctly separated from coxae III-IV, or all coxae equally separated from one another 24

24a   Legs spindly, disproportionately long; body ovoid or nearly circular in outline; gnathosom directed forward whether retracted or protracted; dorsal setae spiniform (p. 201) .... Neophyllobidae

24b   Legs not spindly; longest legs not longer than length of idiosoma, normally shorter 25

25a Dorsal plating normally extensive, if not extensive, stigmata or peritremes not obvious; palp tibia with a pronounced claw; chelicerae generally independently movable (p. 175) .... **Stigmaeidae**

25b Dorsal plating absent or feebly developed, chelicerae fused with each other in midline to form a conical stylophore which bears on its dorsal surface a pair of sinuous peritremata (p. 196) .... **Caligonellidae**

26a Movable chelae long, straight, extrusible; empodia absent; larvae normally with only a propodosomal plate; elongate in shape 27

26b Movable chelae short, curved, hinged at base, larvae normally either with two dorsal plates, or if only a propodosomal plate, seta on palpal coxa posterior to palpal femur 28

27a Gnathosoma (palpi plus mouth-cone) of adult and nymph attached to propodosoma by an extensible collar, capable of being completely withdrawn into collar; with ossiform crista or scutellate shield, body setae flattened and serrate .... **Smaridiidae**

The members of the family Smaridiidae have an elongate, oval body which is pointed anteriorly. There is a crista metopica present or absent; the propodosoma is usually extended into a cone with a shallow furrow separating it from the hysterosoma. The larvae are parasitic on small animals and the adults are predators. The family is little known in North America.

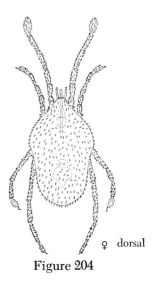

♀ dorsal

Figure 204

Figure 204 *Fessonia lappacea* Southcott

27b Gnathosoma of adult and nymph fixed to front of propodosoma, without extensible collar enabling forward projection of gnathosoma .... **Erythraeidae**

The Erythraeidae are fairly large mites normally colored red or reddish brown and are predators of other animals. The larvae of this family are in part parasitic on certain insects, others tend to be like the adults and feed on plant parts and possibly other arthropods.

Members of this family occur in the mulch layer or on the plants themselves. The species of the genus *Balaustium* has been recorded as attacking people and causing irritation of the skin. This family is fairly large, complex and contains an anterodorsal crista metopica. The outstanding characteristic for the family is the retractile cheliceral digit into a cheliceral sheath, with large stigmatal openings between the cheliceral bases.

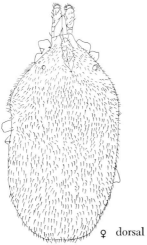

♀ dorsal

Figure 205

♀ ventral

Figure 206

Figure 205 and 206 *Balaustium* sp.

28a Dorsal setae of adult and larvae simple, each located on an individual platelet; two pairs (rarely one) of propodosomal sensory setae, parasgenital scleroties of adult well-developed and with many setae (p. 170) .... Johnstonianidae

28b Dorsal setae of adult and larvae complex; if not, always in association with only a single pair of propodosomal sensory setae 29

29a Tectum with one or two setae; adults may be "figure-eight" shaped or similar to all other mites of this group; larvae with a single dorsal shield; in most species, seta on palpal coxa of larva posterior to palpal femur; urstigmata always associated with coxae I (p. 174) .... Trombiculidae

29b Tectum with numerous setae; adults never "figure-eight" shaped; larvae normally with more than one dorsal shield; seta on palpal coxa of larva normally extending anterior to base of palpal femur; urstigmata associated either with coxae I or coxae II .... Trombidiidae

The family Trombidiidae is one of the largest and most complex groups of prostigmatid mites. They are parasitic in the larval stage and predaceous on other animals as adults. They contain an anterodorsal crista metopica which is normally associated with 1-2 pairs of sensory setae. The cheliceral digit is short and hook-like, unlike the erythraeids this digit is nonretractable.

30a Aquatic mites 53

30b Non-aquatic forms, may be rarely found in water but not capable of swimming and living in water 31

31a Idiosoma entirely encased in thick, reticulated and porous armor having no transverse suture, but with a suture separating anterior and posterior plating;

gnathosoma retractable beneath a prominent sclerotized hood ....
.... **Cryptognathidae**

The members of this family are very distinctive and are a rewarding change to a worker who has been struggling with other mites belonging to the superfamily Raphignathoidea. Slide-mounted specimens are easily recognized. To date there is only a single genus, thus placement is simplified. According to Summers and Chaudhri (1965) knowledge of the habits and distribution is at present scant. It is believed that members of the family are predaceous, however, it is also known that their mouthparts are so loose-jointed and vulnerable to injury that if they are predaceous their prey must be very passive. They are known to occur in low-density in leaf mold and on moss covered substrates. They are known by the author to exist in range pastureland in South Dakota and Texas ....
.... Genus *Cryptognathus*

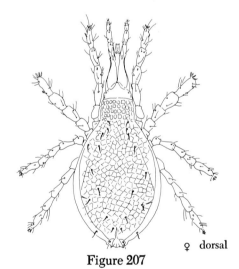

♀ dorsal

Figure 207

Figure 207 *Cryptognathus imbricatus* Summers and Chaudhri

Idiosoma entirely covered with a cuirasslike ornamented exoskeleton of two parts. A vaulted dorsal plate laps onto ventor, infolded and heavily sutured around dorsopleural line to a unified ventral plate. Prosternal apron comprises a crescentic, transparent flange on front margin of ventral plate, three pairs of paragenital setae, peglike sensillum "k" present on genu II, two proximoventral setae on tarsi III and IV.

Collected in California from leaf mold, *Quercus agrifolia*, oak and pine leaf mold.

31b Idiosoma may have plates but these never entirely cover venter; gnathosoma not retractable beneath a hood 32

32a Palps greatly extended, elongate, slender; thumb-claw when present small or obsolete; two pairs empodial raylets; coxae II-III well separated (p. 194) .... **Eupalopsellidae**

32b Without above combination of characters 33

33a Rodlike sensory setae of tarsus I lying flush with tarsus within a specialized membranous depression; propodosoma apex with a tubercle bearing a pair of setae 34

33b Rodlike sensory setae, if present, not lying flush with tarsus in a specialized membranous depression, but rather erect arising from a small circular membranous base; anterior tubercle may be present or absent 37

34a Cheliceral shears large, well developed; palp with long terminal setae; propodosoma with well developed vertical setae .... **Rhagidiidae**

The family Rhagidiidae are inhabitants of caves and are known to be a part of the prairie grasslands fauna of the central United States. They were recorded as common below 10 cm in the soil at the Pawnee I.B.P. Grassland biome site in Colorado. Two species, *Rhagidia weyerensis* (Packard) and *Rhagidia cavernarum* (Packard) are known from Virginia, Kentucky, New York, Tennessee, New Hampshire and Washington D.C. Both species are considered to be obligatory cavernicoles.

34b Cheliceral shears small, sometimes distorted **35**

35a Gnathosoma covered with a rooflike extension, body sclerotized in a tuberculate pattern, idiosoma often with a "V" or "Y" suture dorsally .... **Penthalodidae**

This family is related to mites of the families Eupodidae and Penthaleidae. They are often colorful with fine reticulation both dorsally and ventrally.

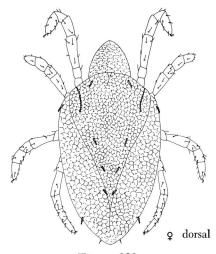

♀ dorsal

Figure 208

Figure 208 *Penthalodes oregonensis* Baker

Rounded shield over the rostrum, dorsum of body with V-shaped indentation; skin pattern of tubercles forming connecting hexagons, covering dorsum and venter; no reticulate pattern on shield over rostrum; anterior median tubercle on propodosoma with a pair of simple setae, nine pairs of pilose genital setae; tarsus I with three rodlike sensory organs.

Collected from Oregon in fir duff.

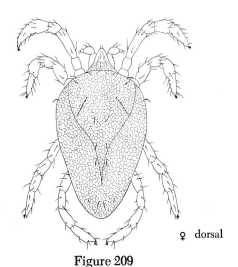

♀ dorsal

Figure 209

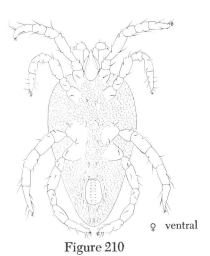

Figure 210

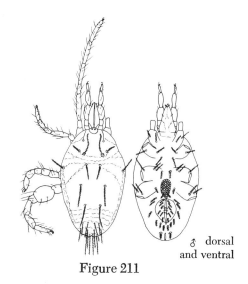

Figure 211

Figures 209 and 210   *Penthalodes turneri* Baker

Reticulate pattern covering dorsum and entire venter, dorsal body setae semiplumose with three to five lateral branches; anterior dorsal setae six-rayed; propodosomatic sensory setae pilose on distal half; genital plate with nine pairs of pilose setae; palpi long, slender.

Collected from Missouri, Texas and South Carolina from peach orchard soil, cotton fields, and soils.

35b   Gnathosoma without rooflike cover; body soft **36**

36a   Anal opening ventral; femur IV often enlarged .... **Eupodidae**

Members of this family are medium-sized mites. Some are associated with mushroom beds and have caused economic damage because of their destruction of the major root system of the mushroom.

Figure 211   *Eupodes* sp.

36b   Anal opening dorsal or terminal; femur IV never enlarged .... **Penthaleidae**

This family contains the well-known winter grain mite, *Penthaleus major* (Duges). It has been recorded as a pest of fall-planted small grains and can cause large losses to farmers. This species is also a pest of peas in California, and has been reported to feed on clover, wild mustard, and lupine. It has a worldwide distribution particularly throughout the temperate zones. During heavy infestations the grain leaves turn brown and die. If the infestation goes unchecked the entire plant dies. Even if checked, the grain becomes stunted and produces little forage or grain. This mite is unique in that cold rather than warm temperatures favor their development. Oviposition is heaviest betwen 50° and 60°F., temperatures between 45° and 50° were found to be optimum for hatching.

Penthaleus major (Duges)

Anal aperature located on the dorsum, surrounded by a reddish-orange spot, body sparsely covered with small setae, chelicerae short, with two teeth on distal end; pedipalps four-jointed, short; legs I and IV longer than legs II and III, tarsi small with two claws and empodium, tibia with a spine on inside near tarsal joint; genital opening ventral, with two pairs of suckers.

Collected from Texas, Oklahoma, Kansas, Missouri, Arizona, California, Tennessee, District of Columbia, Pennsylvania, Massachusetts, New York, Minnesota and South Dakota on small grains, clover, wild mustard, and lupine and on ground under stones.

37a Chelicerae whiplike long, genital opening transverse 38

37b Chelicerae either opposed or scissorlike, if appearing whiplike, then the genital opening is longitudinal 39

38a Chelicerae arising from an eversible stylophore, tarsal claws with tenent hairs, adults occasionally with only three pairs of legs; plant feeders (p. 165) .... Tenuipalpidae

38b Chelicerae not arising from an eversible stylophore, tarsi I, without claws; tarsal claws II-IV without tenent hairs; adults always with full complement of legs; parasites of cockroaches .... Iolinidae

This family contains only the single species which feeds upon the large tropical cockroach, *Blaberus craniifer*. This large cockroach is often utilized as a laboratory research animal. .... Genus *Iolina*

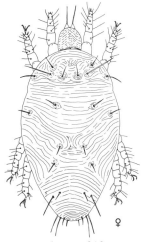

Figure 212

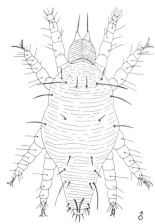

Figure 213

Figures 212 and 213   *Iolina nana* Pritchard

Soft-bodied, movable cheliceral digit long, whiplike; palp simple, one-segmented; stigmatal openings at base of chelicerae; peritremes indistinct; eyes absent; apoteles absent on legs I, genital opening transverse; first and third dorsal propodosomals, dorsocentral hysterosomals, and first lateral hysterosomals short, broadened proximally, attenuate, nude; female opisthosoma with four pairs of pregenital setae and two pairs of paraanals; genital opening a simple transverse slit, anus separate and located near caudal margin; male opisthosoma with two pairs of setae anteroventrally and with four pairs of genital-anal setae; intromittent organ a pair of slender genital stylets.

Collected from Massachusetts on *Blaberus craniifer* and from *Diplotera dystiscoides* imported from Hawaii.

39a Cheliceral bases not fused, the chelicerae moving scissorlike over gnathosoma; anterior portion of gnathosoma extended into a long snout, with two pairs of long sensory setae or sensille on the propodosoma **40**

39b Cheliceral bases fused, not moving scissorlike in motion over gnathosoma **41**

40a Palpi long elbowlike with distal setae; with three pairs of genital discs ....
.... **Bdellidae**

Spinibdella cronini (Baker and Balock)

Mites of the family Bdellidae are predaceous on small arthropods and their eggs. They are very fast moving animals, and occur in all terrestrial habitats where the food material is found. According to Atyeo (1960) they have been recorded from all major land masses of the world and many of the insular groups. These mites are red, reddish-brown or green and are referred to as snout mites because of the gnathosoma consisting of the elongate chelicerae, hypostome and palpi. Because of this snoutlike appearance they are easily recognized as a family group. They are only to be confused with possibly members of the family Cunaxidae.

Bdellid mites vary in size ranging from large to very small. They range in habitat from the intertidal region of coastal shores to pasture soils and plants of the forests. They are commonly found in stored food products. In the United States they have a broad temperature and humidity tolerance; a single species may occur throughout the United States. One species, according to Atyeo (1960), occurs in North America from Mexico to Alaska to Iceland and Europe. Many species have been brought back from arctic expeditions.

40b Palpi long or short without distal setae, approximately equal to chelae in length, distal segments adapted for grasping ....
.... **Cunaxidae**

The family Cunaxidae prey on other animals. They are moderate sized with elongate pyriform bodies. The integument is soft, and finely striated, with a suture between the propodosoma and hysterosoma. The gnathosoma is long and conelike with raptorial palpi, and elongate chelicerae. They are often brightly colored red or reddish-brown and normally associated with leaf litter or mulch.

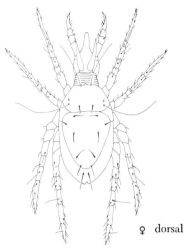

♀ dorsal

Figure 214

Figure 214 *Cunaxa capreolus* (Berlese)

Dorsal scutum transversely divided into two scuta; palpus without a mesal process between tibia, and tarsus but with a distinct femoral process which is small and spurlike.

Collected in Florida and Texas.

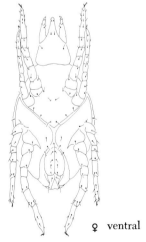

♀ ventral

Figure 215

Figure 215 *Neocunaxoides pectinellus* (Muma)

Tarsus of the palpus with one large process, dorsum of scutum with spines, punctuate, and undivided.

Collected from Florida, Texas and South Dakota.

41a **Chelicerae and rostrum fused into a cone; parasites of vertebrates or arthropods** **42**

41b **Chelicerae and rostrum not fused into a cone; free-living** **46**

42a **Palpal tarsus clawlike, with small lobe possibly a remnant of a tarsal thumb, tarsus of leg I without claws or empodium, tarsus of legs II-IV, with large disc-shaped empodia; ectoparasites of arthropods** .... **Heterocheylidae**

Members of this family were formerly classified in the family Pseudocheylidae by Tragardh (1950). *Heterocheylus* sp. has been collected beneath the elytra of a passalid beetle, *Popilius disjunctus* from Georgia. Krantz (1970) records it as collected with regularity in the southeastern United States.

42b **Palpal tarsus not clawlike, without a small lobe, tarsi of other legs variable but not as stated above** **43**

43a **Legs I reduced, adapted for clasping hairs of host; legs II-IV normal with one or two claws; parasites of mammals** .... **Myobiidae**

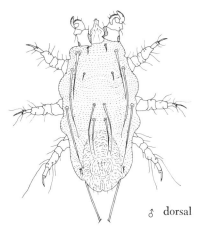

♂ dorsal

Figure 216

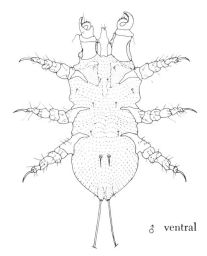

♂ ventral

Figure 217

Figures 216 and 217   *Blarinobia cryptotis*
McDaniel

Myobiids are parasitic mites adapted to live in association with the fur of certain mammals. The front legs are modified to clasp the hair of their host. Myobiid mites are encountered in the United States on mammals of the orders Insectivora, Chiroptera, Rodentia and Marsupialia.

Members of this family have a diverse feeding habit. Some species feed and are confined to the hair follicle base and, in most infestations, do not cause hematophagy of the infested site. However, other species are known to be blood feeders and can cause loss of hair and dermatitis. Myobiid mites are obligate parasites thus spending their entire life on the host. The only movement is from host to host or from one region of the host body to another. Some species are restricted to definite host body regions.

Much of the work on this group has been directed toward myobiids found on bats. Ewing (1938) first treated the United States fauna in his work *"North American Mites of the Subfamily Myobiinae."* Workers in this group should consult the works of Jameson (1955), Dusbabek (1969) for recent revision of the genera of myobiids associated with mammals.

43b   Legs I normal, if reduced, not adapted for clasping hairs, used for movement or modified other than described above
.... **44**

44a   With large hooklike spine on venter of femora of legs I-IV; palpal segments telescoped ventrally; empodia pad-like; skin parasites of mammals ....
.... **Psorergatidae**

The species that belong to the family Psorergatidae are very small, round mites that infest the skin of mammals. Because of their small size, they are commonly found only when the host has a dermatitis condition.

44b   Without large hooklike spine on venter of femora of legs I-V; palpal segments normal or not as described above **45**

45a Empodia simple, with a double row of tenent hairs, body elongate, claws present; associated with the quills of birds (p. 202) .... Syringophilidae

45b Empodia cuplike with rayed projection; body rounded; claws reduced or lacking; parasites of snakes .... Ophioptidae

Members of this family are known as pit mites of snakes. They have been found on South American snakes in zoos in the United States. As the name implies, these mites produce and live in small pits in the heavily cornified layers of snake scales. Several species of this family have been described from various parts of the world.

In the United States members of this family are introduced with the host snake.

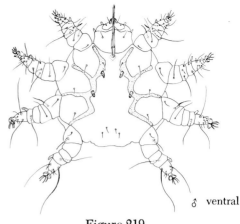

♂ ventral

Figure 219

Figures 218 and 219 *Ophioptes parkeri* Sambon

Collected from *Clelia rustica* (recorded by Sambon, 1928 as *O. oudemansi* from Argentina).

46a Strongly armored; normally bright orange or yellow (color will fade in alcohol); a pair of large lenslike structures laterally on the idiosoma in addition to ocelli; coxae forming apodemal patterns, chelae with opposed digits; free-living .... Labidostommidae

Members of this family are heavily armored ornamented and moderately large, ranging in size from 500 to 2000 microns. They have been collected mostly from humus, lichens, soil and moss. They are predacious, feeding on small invertebrates. .... Genus *Labidostomma*

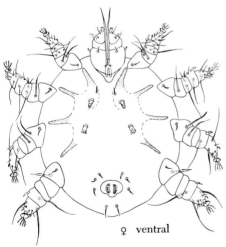

♀ ventral

Figure 218

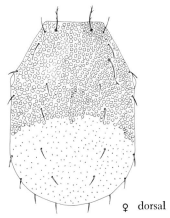

♀ dorsal

Figure 220

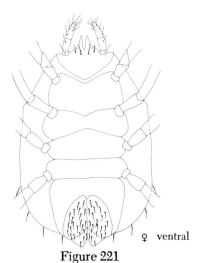

♀ ventral

Figure 221

Figures 220 and 221 *Labidostomma barbae* Greenberg

Single-pored tubercles absent, pair of multi-pored lateral tubercles usually present, dorsum without antero-lateral projections, chelicerae toothed, famulus absent from pre-tarsus of leg I, idiosoma rotund, body setae simple, dorso-central setae subequal, about same length as dorso-lateral setae, anterior and posterior pseudostigmatic bases subequal, both pairs of pseudostigmatis setae subequal,

branched, posterior pair with longer branches, ventral opisthosomal plate parallel-sided, enveloping genital plate in male, only partially enveloping genito-anal plate of female.

Collected from Georgia from leaf mold and moss.

46b   Soft-bodied, not armored; without large lateral lenslike structures or strong apodemal patterns 47

47a   Cheliceral digits opposed; propodosoma with one or two pairs of pseudostigmata and pseudostigmatic organs 48

47b   Cheliceral digits not opposed, fixed digit reduced; movable digit short and needlelike 51

48a   Empodium and two claws present on tarsi I-IV; three pairs of genital discs, fixed chela normal; two pairs of pseudostigmatic organs .... Pachygnathidae

Members of this family are small, round-bodied mites, free-living in areas with a high organic composition. They are said to be related to the Oribatid mites and members of the family Nanorchestidae.

48b   Empodium or true claws absent on all or some tarsi 49

49a   Claws present on all legs, empodia absent; without ocelli; one pair of club-shaped pseudostigmatic organs on the propodosoma .... Pediculochelidae

In this work the placement of the members of Pediculochelidae follow that of Krantz (1970) and the characterization of Pediculochelidae by Price (1972). Pediculocheloid mites are minute, weakly sclerotized, elongate mites;

without eyes; idiosoma with distinct humeral sulcus; hysterosoma divided into four regions by three transverse dorsal sutures; legs I and II separated from legs III and IV by first hysterosomal region; gnathobase with rutella; genital region with two pairs of discs.

♀ dorsal

Figure 222

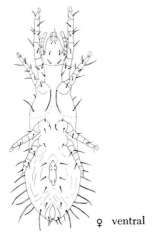

♀ ventral

Figure 223

Figures 222 and 223 *Pediculochelus lavoipi-errei* Price

Some hysterosomal setae with swollen section near base; long setae between coxae II; a pair of setae between coxae IV; one seta on trochanter III, one seta proximal to palp trochanter; two setae on palp femur; one solenidon on tarsus I; propodosoma with middorsal region bearing a pair of vertical setae between chelicerae; hysterosoma I with two pairs of setae in anterior row, two pairs in posterior row; hysterosoma II bearing legs III and IV; hysterosoma III with two pairs setae; hysterosoma IV with two rows of setae, six in each row.

Collected from California in grassland soil.

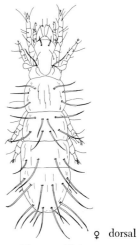

♀ dorsal

Figure 224

♀ ventral

Figure 225

Figures 224 and 225 *Pediculochelus parvulus* Price

Hysterosomal setae without swollen section near base; trochanter III without setae; gnathobase with three pairs of setae; two setae proximal to trochanter on coxal base of palpi; ventral setae between coxae II as long as setae between coxae I.

Collected from California in grassland soil.

49b Claws absent; clawlike empodia present on all legs **50**

50a With two pair of pseudostigmatic organs .... **Nanorchestidae**

This family contains only a few described species. They are very small mites with a worldwide distribution. They are common inhabitants of pasture grasslands in South Dakota.

50b With one pair of pseudostigmatic organs .... **Alicorhagiidae**

This family contains the single genus, *Alicorhagia*, which contains only a few described species from Europe and Japan. Undescribed members of this family have been taken from soil samples in pasture grasslands in South Dakota.

51a Hysterosoma divided transversely by one or two sutures, empodia clawlike; tarsus I with several erect dorsal sensory rods; elongate forms with distinct peritremes on anterior portion of propodosoma; free-living .... **Paratydeidae**

Members of this family are found in the soil and are thought to prey on other minute animals. Krantz (1970) stated that collections of this family have been obtained from the nests of birds and from moss in Oregon. They have also been obtained from South Dakota from pasture grasslands. These recent collections indicate that more members of this family will be found in the United States ....
.... Genus *Neotydeus*

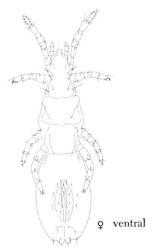

♀ ventral

Figure 226

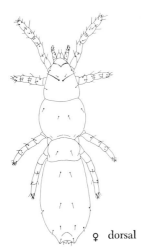

♀ dorsal

Figure 227

Figures 226 and 227 *Neotydeus ardisanneae* Baker

Body divided into four portions by three transverse sutures; palpi four segmental, without thumb-claw complex; hysterosoma divided into three distinct parts by two transverse sutures just behind the posterior coxae; body setae short; propodosoma with a single pair of long sensory setae; two pairs of lateral peg-like setae; two pairs of genital suckers; four pairs of genital and four pairs of accessory setae; without eyes.

Only the single species is presently recorded in this genus.

Collected from Illinois in leaf trash, Texas and South Dakota from pasture soils.

51b   Hysterosoma without transverse sutures; empodia pad- or hair-like tarsus I with an erect sensory rod; peritremes indistinct or absent 52

52a   Legs I-II with shell-like pattern, with an ereynetal organ opening in the distal portion of tibia I, free-living; associated with mollusk or parasitic in nasal passages of vertebrates (p. 189) ....
.... Ereynetidae

52b   Legs I-II without netlike pattern, without an ereynetal organ opening; with padlike empodium, body with soft, striated integument, pseudostigma present; sensory setae club-shaped (p. 183) ....
.... Tydeidae

53   The water mites are unique within the chiefly terrestrial mites and ticks. Most of the members of this group of mites are found to be generally adapted to freshwater. There are a few species that are to be found in brackish and salt water but none of the members are to be found terrestrially.

They are bright colored and are more commonly collected in the northern lakes and ponds with high quantities of rooted aquatic vegetation. They look very much like their relatives, the spiders, being soft bodied with a series of heavily sclerotized plates. The most conspicuous body structure is the legs which are normally long and supplied with spines, setae and long hairs.

More than 90 percent of all published information on American species is based on the works of Robert Wolcott, whose publications extend from 1899-1918, and Ruth Marshall, between 1903-1944.

Identification to genus and species is difficult and based on the type and position of the leg epimera, genital field, capitulum and palps. They are also quite difficult to make slide mounts, with most of the species requiring an incision in the dorsal body after which the specimen is placed in potassium hydroxide to render it sufficiently translucent for work under a compound microscope.

Family Tarsonemidae

The mite family Tarsonemidae is associated with a group of families that exhibit a tendency toward loss of some of the legs or toward reduction or enlargement in size of the legs. In the Tarsonemidae, the last pair of legs in females is greatly reduced and terminate in a long whiplike setae. In the males the last pair of legs is greatly modified to function as accessory copulatory appendages.

Members of this family are of importance to agriculture. The cyclamen mite, *Steneotarsonemus pallidus* Banks, causes extensive damage to numerous greenhouse crops. Some of the species that are considered to be of economic significance include: *Tarsonemus randsi,* a pest of considerable importance to commercial mushroom production and the maintenance of fungus cultures in research laboratories; *Steneotarsonemus spirifex* which attacks corn, oats and other grains; *S. bancrofti* which feeds on newly planted stocks of strawberries and sugar cane cuttings; *S. pallidus,* causes the greatest amount of damage to plants; and *Hemitarsonemus latus* which is second only to *S. pallidus* in importance of the latter two forms. Both have enormous host ranges which include many of our commercially grown crop plants.

Members of the genus *Iponemus* are associated with the economically important bark beetles. The mites are egg parasites of pine bark beetles. Members of this genus (Lindquist, 1969) are monospecific and were ranked in order of decreasing size, based on size of unengorged female and size of host beetle.

Members of this family are phytophagous, fungivorous, insectophilous and parasitic in food habits. Members of the genus *Steneotarsonemus* tend to feed on monocotyledonous plants. *S. ananas* is restricted to pineapple, *S. latipes* may feed on banana, and *S. laticeps* feeds only on plants of the family Amaryllidaceae. However, a few members of the genus are not restricted and feed on dicotyledonous plants or fungi. All species of the genus *Tarsonemus* are considered to be fungivorous in habit. *Steneotarsonemus pallidus* and *Hemitarsonemus latus* have a wide range of host food plants.

In the past many workers have suspected that tarsonemid mites might be important in the field of health and medical acarology. However, no conclusive evidence has been brought forth to date which would incriminate the members of this family in the transmission of human disease.

Tarsonemids are very small mites, with the body divided into well-defined portions. There is pronounced sexual dimorphism, the males being much smaller than the female. Both sexes are dorsoventrally depressed which aids them in living between the sheaths and stems of grass hosts and the integument of insects. They are characterized by a pronounced development of apodemes on the ventral portion of the body. There is no evidence of a tracheal system in the males. Classification of members of this family is based almost exclusively upon the structure of the hind pair of legs of the male; therefore it is essential for a collector to obtain the male in his collecting of material. Keys to species are based upon the male.

1   **Males with inner flange on femur IV absent or greatly reduced in size; fourth dorsal propodosomal seta always laterad from third setae; capitulum longer than broad, subcordate; palpi never prolonged to form snoutlike beak; female with lobelike tracheal expansions absent or reduced in size, never bilobed; first pair of ventral propodosomal setae never in front of apodemes I** .... **Genus *Tarsonemus***

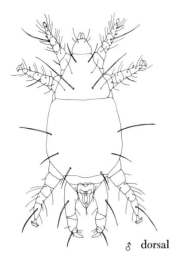

Figure 228

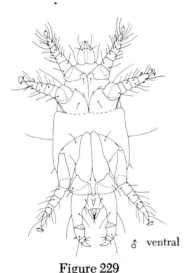

Figure 229

Figures 228 and 229   *Tarsonemus randsi* Ewing

Male femur IV more than twice as long as broad at base; tactile seta of tibia IV about two-thirds as long as femur IV, femur IV not angulate at base, tactile setae of tibia IV never as long as leg IV, usually shorter than femur IV, femur IV without inner flange; third dorsal propodosomal setae long, much longer than setae of other three pairs; femur

IV broadest near base, width at mid-segment much less than half length of segment; tactile seta of tibia IV exceeded in length by at least one other seta on leg; female first ventral propodosomal seta located on or in front of apodeme I; transverse apodeme uninterrupted; hysterosoma without a pair of genital setae situated between coxae IV.

2   **Palpi of both sexes prolonged anteriorly forming an elongation of the capitulum .... Genus *Rhynchotarsonemus***

Rhynchotarsonemus niger Beer

Both sexes jet black, body broadly oval, broadest at middles of coxae III, tapering abruptly caudad; dorsum of propodosoma projected anteriorly forming a broad cephalothoracic shield with a truncate anterior margin as broad as capitulum.

Collected from California from *Quercus lobata* infested with *Asterolecanium minus*, lemon fruit, *Hydrangea quercifolia*, under bark of *Hydrangea quercifolia*.

3   **Males with large, flangelike expansion on inner margin of femur IV or if absent then fourth dorsal propodosomal seta in linear arrangement with setae of three preceding pairs; capitulum broader than long; females never with conspicuous transverse apodeme near main body suture .... Genus *Steneotarsonemus***

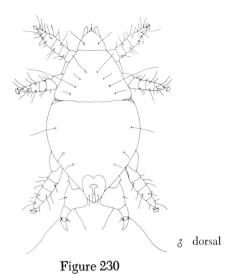

$\male$ dorsal

Figure 230

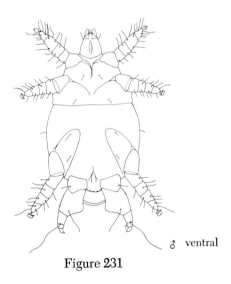

$\male$ ventral

Figure 231

Figures 230 and 231 *Steneotaronemus pallidus* (Banks)

Male tactile seta of tibia IV longer than femur IV; fourth dorsal propodosomal setae shorter than setae of remaining three pairs and not in linear arrangement with the other setae, second dorsal propodosomal nearly as long or longer than other dorsal propodosomal; posterior median apodeme bifurcate caudally, arms extending anteriorly from inner basal angles of coxae IV, anterior median apodeme extending indistinctly from main body suture to point opposite posteromedial termina of apodemes II, well defined from this point to its anterior extremity of V-shaped juncture of apodemes.

Collected from California, from cyclamen (greenhouse), ivy (greenhouse), field grown strawberries.

4   **Tarsi II and III without claws, with well defined bell-shaped empodia, tarsi I with claws; leg IV with five segments; femur IV with spurlike process on inner margin; tarsus IV terminating in knoblike claw .... Genus *Neotarsonemus***

Male leg IV with terminal claw reduced to a small tubercle; apodemes I slightly longer than half the greatest width of capitulum; propodosoma with four pairs of setae, first pair shorter than others, second pair longest, third and fourth pairs subequal in length; hysterosoma with five pairs of dorsal setae, and three pairs of ventral setae; leg IV with coxa rectangular, as broad as long, with one stout setae; tibiotarsus surmounted by small blunt buttonlike claw.

Collected from Washington, D.C., California, from shoots of mango (greenhouse), leaves of *Vigna sesquapedata* (greenhouse.)

5   **Males with body laterally compressed; tibia and tarsus or tibiotarsus IV slender, elongate, more than three times as long as basal width of tibia or tibiotarsus, without inner flangelike process on femur IV; female with cephalothoracic shield projected anteriorly covering most of capitulum ....**
.... Genus *Hemitarsonemus*

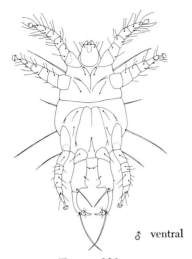

♂ ventral

Figure 232

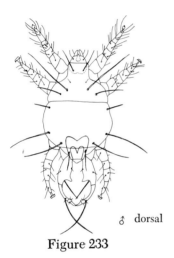

♂ dorsal

Figure 233

Figures 232 and 233 *Hemitarsonemus tepidariorum* (Warburton)

Male with a subapical spurlike process on the inner margin of femur IV, coxa of leg IV narrow, subquadrangular, longer than broad, one ventral seta, femur twice as long as broad, basal two-thirds of segment with broad inner margin terminating in a spur; propodosoma with four pairs of dorsal setae, anterior pair of setae long, first ventral propodosomal setae small; apodemes well developed, conspicuous,

those of leg I as long as genu I extending posteromedially from anterior basal angles of coxae I to juncture with anterior extremity of median apodeme.

Collected in Minnesota, California, from ferns, *Pteris* sp., *Asplenium bulbiferum* and *Polystichum* sp.

6   **Males without claws, empodia or knob-like arolia on tips of legs IV; three segmented coxa, femur and tibiotarsus, tibiotarsus expanded bulbously at tip ....**
.... **Genus *Xenotarsonemus***

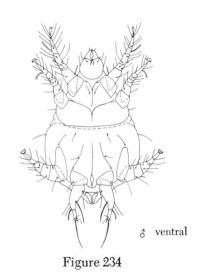

♂ ventral

Figure 234

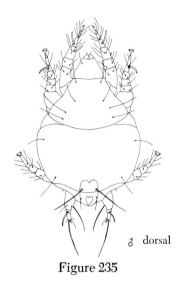

♂ dorsal

Figure 235

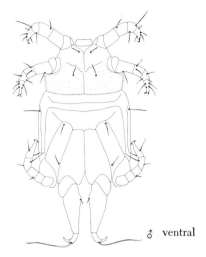

♂ ventral

Figure 236

Figures 234 and 235 *Xenotarsonemus viridis* (Ewing)

Male with the characters of the genus, legs rather long and stout; a conspicuous hysterosomal suture transects body dorsally at anterior two-fifths of genital papilla; dorsal propodosomal plate extended forward to form a broad cephalothoracic hood; apodemes I slightly longer than genu I extending in posteromedial directions from anterior basal angles of coxae I, converging medially to form a "Y"-shaped juncture with anterior extremity of median apodeme.

    Collected in Maryland from strawberry.

7   **Male with dorsal body setae long, some coarse and spiculate; tibia IV much longer than wide; female with dorsum marked off into plates; some dorsal setae enlarged apically; palpi of both sexes long .... Genus *Daidalotarsonemus***

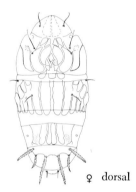

♀ dorsal

Figure 237

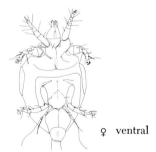

♀ ventral

Figure 238

Figures 236, 237 and 238 *Daidalotarsonemus jamesbakeri* Smiley

Dorsomedian longitudinal plates of hysterosoma contiguous; female dorsum of propodosoma with two pairs of setae, first pair longest, finely scabrous, second pair simple, saberlike; lateral margin of capitulum notched; platelets of propodosoma irregular in shape and size, two platelets punctate, dorsum of hysterosoma with six pairs of setae; male, hysterosoma with two pairs of setae located above and below suture, apodeme I shorter than apodeme II; apodeme III and IV longer than anterior median apodeme; a pair of lateral plates posterior to propodosomal and hysterosomal suture.

Collected from North Carolina from blueberry buds.

8   Dorsal hysterosomal setae barbulate, long, some as long as or longer than width of body at hysterosomal suture; palpi moderately long; female with secretions of wax on dorsum; male with femur IV elongated; tibia IV about as wide as long.... Genus *Ceratotarsonemus*

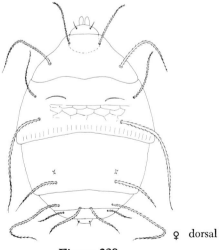

♀ dorsal

Figure 239

Figure 239 *Ceratotarsonemus scitus* DeLeon

With the characters of the genus.
Collected in Florida, from *Litchi chinensis.*

9   Male heteromorphic; propodosomal and hysterosomal setae serrated; venter of propodosoma and hysterosoma, each with two pairs of simple setae; leg IV is robust with femur having a triangular shaped flange; female pretarsi I with vestigial claw, pretarsi II and III each with only a single outer lateral claw; leg I tarsus and tibia fused; femur II with ventral apophysis ....
.... Genus *Heterotarsonemus*

♂ ventral

Figure 240

Figure 240 *Heterotarsonemus lindquisti* Smiley

With the characters of the genus.
Collected from Louisiana from inner bark of loblolly pine with *Dendroctonus frontalis.*

10   Female, obligate parasites of the egg of ipine Scolytidae, beetles which undergo

a hysterosomal physogastry during the parasitic feeding period; pharyngeal pump conspicuously enlarged, its greatest width equal to about half or more of the greatest width of the gnathosoma, pharyngeal pump without strongly sclerotized loop .... Genus *Iponemus*

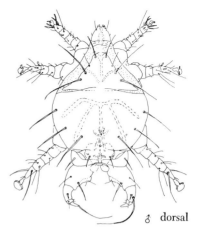

♂ dorsal

Figure 243

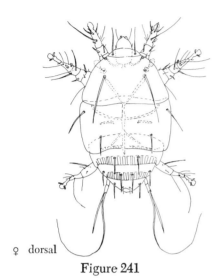

♀ dorsal

Figure 241

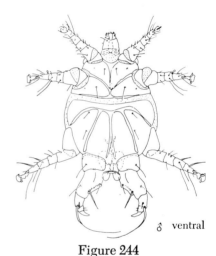

♂ ventral

Figure 244

Figures 241, 242, 243, and 244 *Iponemus integri* Lindquist

This genus contains a group of tarsonemid mites that are egg parasites of ipine bark beetles, they are host specific with these beetles. A worker of this group should consult the works of Lindquist (1969), Lindquist and Bedard (1961) in regard to the classification of this very special group of mites.

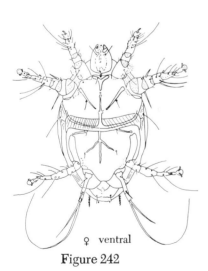

♀ ventral

Figure 242

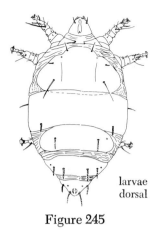

larvae
dorsal

Figure 245

larvae
ventral

Figure 246

Figures 245 and 246 *Iponemus integri* larvae

Family Tetranychidae

The family Tetranychidae, prostigmatic mites, has specialized mouthparts for piercing plant cells and sucking out their contents. Most members feed on the foliage of plants. The removal of the content of the foliage cells results in the appearance of white patches on leaves. These white regions later turn brown and the leaf, when heavily infested, tends to curl.

This is one of the most important families within the Acari, since a large number of the species are pests of many of our agricultural crops. It has been established that almost all of the major food crops and ornamental plants have one or more species that utilize them as a source of food. Some of the mites of this family have the ability to spin a silken webbing that covers the foliage of the parasitized plant and the mites live and lay their eggs beneath this webbing.

The Tetranychidae are found in all parts of the United States and have a worldwide distribution. The color of this group has long been used in describing many of the economic species. The color ranges from green to yellow-green to red and orange. One of the common species encountered by the homeowner is *Bryobia praetiosa*, the clover mite, which is a common parasite of lawns. In the fall this species invades houses in great numbers and causes a home owner much concern due to its presence.

Three species of the family Tetranychidae have long played an important role in damaging important food crops throughout the world. These are the Green two-spotted mite *Tetranychus urticae* (Koch), the Linden mite *Eotetranychus tilliarium* (Herman) and the Carmine mite *Tetranychus telarus* (L.).

One who undertakes the task of working with this group of mites should familiarize himself with the works of Pritchard and Baker (1955), Jeppson, Keifer and Baker (1975), Tuttle and Baker (1968). No attempt is made to treat the more than 130 species occurring in the United States.

Empodium split distally, with free proximoventral hairs; duplex setae of tarsus I widely separate, dividing segment into 3 equal parts, a single pair of para-anal setae; aedeagus always bends dorsally, setae not borne on tubercles in either sex .... **Genus *Tetranychus***

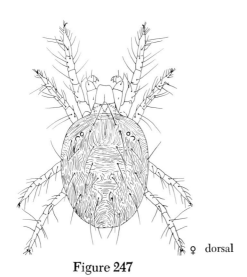

♀ dorsal

Figure 247

Figure 247 *Tetranychus telarius* (Linnaeus)

Male aedeagus with axis of knob parallel or forming a small angle with axis of shaft, posterior angulation no longer than anterior angulation; female with longitudinal striae between the third pair of dorsocentral hysterosomal and with a diamond-shaped pattern in the area caudad of these setae; empodium with the empodial spur very small or absent; tarsus I with proximal pair of duplex setae distad of other proximal tactile setae; peritreme with distal end forming a simple hook; empodium with 3 pairs of empodial hairs, all similar in length; mediodorsal spur of empodium not over one-third as long as proximoventral hairs.

This is the well-known carmine spider mite of North America. It has been referred to as *T. cinnabarinus* by many workers until the work of Boudreaux and Dosse (1963) which firmly established the name *T. telarius*. *T. telarius* is widespread throughout the United States on all types of plant hosts.

Family Pyemotidae

The family Pyemotidae includes some mites of medical or agricultural importance. Presently only a small fraction of the fauna has been collected and described. Estimates of the number of species has ranged as high as 120. The fragmented representation of the total fauna, the scarcity of specimens and their small size makes definition of species and higher categories very difficult. Many members, such as those associated with grain (*Pyemotes sp.* and *Acarophenax tribolii*) have widespread distribution. However, many forms within the family have restricted distribution.

Gravid adult females may become greatly distended and spherical. However,

Pyemotes ventricosus, the hay itch mite, and others, maintain their normal body shape. Most of the classification within this family has been based on females. The number of collected males is small as they are considered to be parasitic upon the body of the female. The food habits of members of the family Pyemotidae range from being fungivorous, parasitic on insects and their eggs, to scavengers and phytophagous.

The hay itch mite, a parasite of the Angoumois grain moth and other grain inhabiting insects, will attack people who come into contact with grain, straw, hay, grasses, beans, peas, cottonseed, tobacco or broomcorn that is infested by the above insect larvae. As many as 200-300 bites per person have been reported causing fever, malaise, vomiting, backache and secondary infection to the victims.

Phytophagous or saprophagous forms are found in the genus *Siteroptes*. *Siteroptes cerealium*, is a principal vector of *Fusarium poae* known as silver top of grasses and central bud rot of carnations, it also has been implicated in the transmission of *Nigrospira oryzae*, a fungal parasite of grains.

Members of the genus *Acarophenax*, *A. tribolii*, are found associated with *Tribolium confusium*, the confused flour beetle. The mite pierces the cuticle of the beetle with stylet-shaped chelicerae and removes the body contents. The genus *Pygmephorus*, has many members associated with the nests of rodents and insectivores. The species *Peponocara cathistes* is a parasite of the cockroach, *Arenivaga apacha*, found in the deserts of the southwestern United States.

Because of the difficulty in working with these forms only selective pyemotids are treated in this work. The work of Cross (1965) forms the basis of the treatment of this family.

1   Tarsus I with two claws; peritremes lacking; palps arising laterally from gnathosoma; chelicerae small, indistinct; propodosoma with two pairs of setae; gnathosoma as broad as or broader than long; male with tibia III without enlarged spinelike seta which reaches tip of tarsus .... **Genus *Dolichocybe***

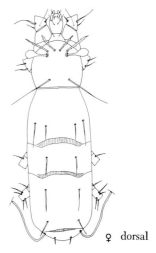

♀ dorsal

Figure 248

♀ ventral

Figure 249

Figures 248 and 249   *Dolichocybe keiferi* Krantz

Pseudostigmata in anterior one-third of body; posterior pseudo-stigmatal setae at least twice as long as either of other pairs; peritremes not visible; anterior ventral plate with six pairs of setae, tarsus I with two claws; chelicerae hooklike; propodosoma with three pairs of setae; gnathosoma narrow and elongate.

Collected from Arizona, California, Washington, from *Acer palmatum,* custard apple, beneath apple bark, in tunnel of wood borer in red bud.

2   **With claws or empodial claw on all legs; tarsus with only an empodial claw; palpi closely appressed to gnathosoma, the latter usually appearing circular in outline; posterior end of idiosoma usually bluntly pointed; tarsus I subequal to or shorter than tibia I; claw of tarsus I never heavy and pincerlike; dorsum of propodosoma always with three pairs of setae .... Genus *Pyemotes***

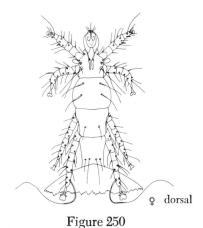

♀ dorsal

Figure 250

Figure 250   *Pyemotes ventricosus* (Newport)

Tracheal system opens on anterior propodosomal region; tarsus I with claw-like empodium, true claws lacking; tarsus II, III, and IV with membranous empodium; idiosomal dilation of female extends anteriorly only to legs IV; hysterosoma more than twice as long as propodosoma, divided into five segments, becoming progressively smaller towards posterior region of body; ventral apodemes of legs I joined to a long sternum, reaching almost to end of propodosoma; apodemes of legs II joined to epimerites I not to sternum; genital opening at extremity of body; femur

of leg IV subdivided into a short basi-femur and a longer telo-femur; legs II and III undivided.

The distribution of this species is considered to be cosmopolitan and can be expected to occur in association with stored grains that are infested with various insect pests upon which they are parasitic. This species of mite may form large populations and attack handlers and carriers of the infested materials. The bite may cause mild to severe cutaneous reactions causing severe itching, nausea, and other symptoms. Feeding infested grain to animals such as chickens, causes death to the animals.

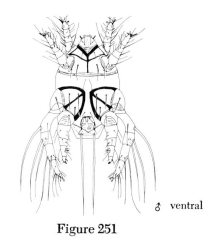

♂ ventral

Figure 251

Figure 251   *Pyemotes parviscolyti*

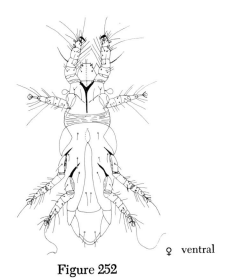

♀ ventral

Figure 252

Figure 252 *Pyemotes parviscolyti*

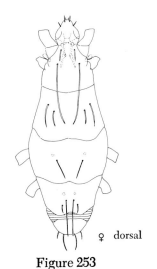

♀ dorsal

Figure 253

Figure 253 *Pymotes* sp.

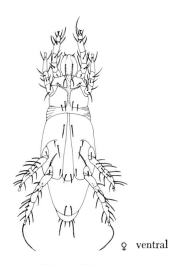

♀ ventral

Figure 254

Figure 254 *Pyemotes* sp

3   **Palpi free distally from gnathosoma, latter appearing elliptical in outline; idiosoma usually truncate or angled behind; tarsus I with only empodial claw, subequal to or shorter than tibia I, claw of tarsus I never heavy and pincerlike; chelicerae not hooklike; dorsum or propodosoma always with three pairs of setae** .... **Genus** *Siteroptes*

Cross (1965) studied nine species including S. *cerealium* and S. *primitivus* from many parts of the United States from the following: Salt marsh grass, *Celtis occidentalis*, carnations, cotton bolls, milomaize, *Setaria* sp., nest of *Microtus* sp. (a mouse), *Fagus grandifolia* in hole of trunk, peach, orchard soil, decaying log and forest floor litter. S. *graminum* is widespread throughout the United States in association with central bud rot of carnations. S. *cerealium* in the principal vector of a disease called "silver top" of meadow grasses. Both diseases are caused by a fungus. The relationship between the fungus and the mite is stated to be a symbiosis in which the mite disseminates and innoculates the fungus which alters the plant tissue to provide a more suitable food source for the mite. The mites are also known to feed on the fungus.

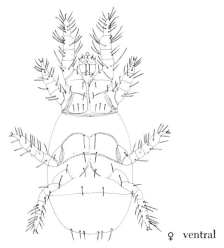

Figure 255

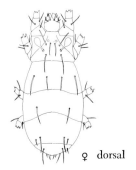

Figure 256

Figures 255 and 256   *Siteroptes absidatus* Cross

Tarsus I with two clavate solendia, solendium three rodlike, curved, slightly longer than claw, arising at level of pedicelar base, solenidium four rodlike, straight, less than half as long or wide as solenidium three arising be-

hind three but in apical third segment; peritremes widely spaced.

Collected from Arkansas.

Cross (1965) states in his treatment of the family Pyemotidae that in addition to *S. absidatus*, six undescribed species are known by him from many parts of the United States associated with tree holes in *Fagus grandifolia*, stems of *Paeonia* sp., corm of *Gladiolus* sp., oats and in cattle droppings.

4   **Gnathosoma bearing free palpi and dorsal, approximate stigmata, in that coxae IV are quadrate; first hysterosomal segment is tripartite; female trochanter I bears five setae; leg I of male clawless** .... **Genus *Trochometridium***

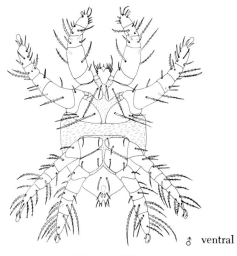

Figure 257

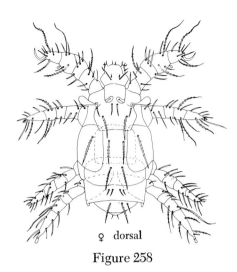

♀ dorsal

Figure 258

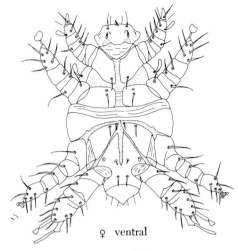

♀ ventral

Figure 259

Figures 257 and 258 *Trochometridium tribulatum* Cross

With the characters of the genus. *T. tribulatum* is the only species in the genus *Trochometridium*. Its distribution is Utah, Kansas, Maryland, Oregon, Iowa, Nebraska, and Texas. Hosts include, cells of *Halictus farinosus Sphecodes arvensiformis*, cells of *Calliopsis andreniformis*, Mutillid, cells of *Nomia melanderi, Nomia bakeri*. To date the known species are parasites of larval bees.

5   Legs I five-segmented; gnathosoma free, at least twice as wide as long; palps fused; body short, fusiform; opisthosoma very short, projecting behind posterior margins of coxae IV only about as much as gnathosoma projects forward from anterior margins of coxae I (figs. 259 & 260) .... Genus *Caraboacarus*

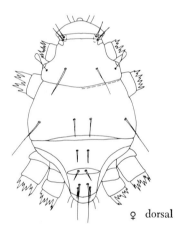

♀ dorsal

Figure 260

Figures 259 and 260 *Caraboacarus sp.*

Collected from Kansas and Michigan, from members of the beetle family Carabidae.

6   Legs II to IV with claws; legs I with four segments; pseudostigmata absent; anterior ventral plate with only two pairs of setae; gnathosoma concealed within the propodosoma, chelicerae stout, easily visible; peritremes opening dorsally, dorsal propodosomal U-shaped

apodeme lacking ....
.... **Genus *Paracarophenax***

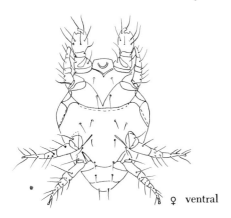

Figure 261

Figure 261 *Paracarophenax dybasi* Cross

Posterior margin of posterior ventral plate distinctly tripartite; dorsal setae except laterals IV stout, sparsely plumose; tarsus II and tarsus III possess two and one spinelike setae respectively.

Collected from Florida from berlese sample of decaying fruit.

7 Leg IV without claws; peritremes opening anteriorly, dorsal propodosomal U-shaped, apodeme present; tibio tarsus I with stout, sessile claw; posterior ventral plate with five pairs of setae ....
.... **Genus *Acarophenax***

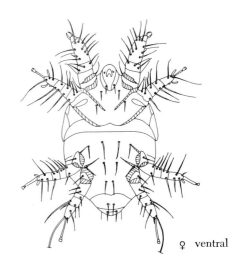

♀ ventral

Figure 262

Figure 262 *Acarophenax tribolii* Newstead and Duvall

This species is associated with stored grain beetles such as *Tribolium confusum* and other grain-inhabiting forms. The figure included herein is adapted from Cross (1965) which was collected from India (taken at Puerto Rico in quarantine).

8 Tibiotarsus I clawless; posterior ventral plate with three or four pairs of setae; small mites, oval; leg I four-segmented; anterior ventral plate with one or two pairs of setae; pseudostigmata lacking
.... **Genus *Adactylidium***

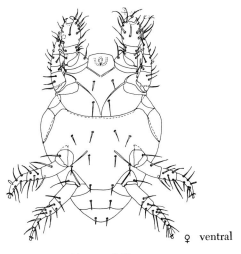

Figure 263

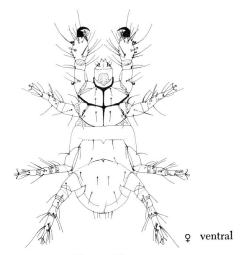

♀ ventral

Figure 264

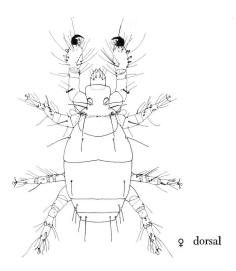

♀ dorsal

Figure 265

Figure 263 *Adactylidium beeri* Cross

Solenidium of tibiotarsus I clavate located in apical half of segment; anterior setae of anterior ventral plate not visible, perhaps lacking, setae of segment III are subequally spaced.

Collected from Michigan, from thorax of Thysanoptera.

9 Seta "c' of trochanter I straight and stout, obtuse or spatulate apically; leg I with four segments, claw I sessile, often deformed; propodosoma with three pairs of dorsals .... Genus *Pediculaster*

Collected from District of Columbia from a fly, *Platycnemis imperfecta*.

10 Propodosoma with one or two pairs of setae; tibiotarsus I with little or no apparent swelling, never attaining twice the width of genu I, claw I and IV present; dorsum of body lacking scalelike integumental engraving (figs. 264 & 265) .... Genus *Pygmephorus*

Figures 264 and 265 *Pygmephorus brachycercus*

Associated with the insect families Scarabaeidae and Buprestidae. A member of the genus is included but no attempt has been made to treat species of this group.

11 Apodemes IV never reaching more than four-fifths of distance to margins of

coxal foramina III; opisthosomal sternum lack engravings ....
.... Genus *Microdispus*

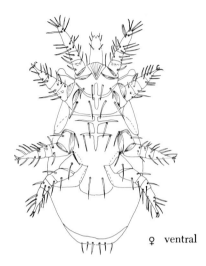

Figure 266

Figure 266   *Microdispus chandleri* Cross

Laterals IV spaced two-thirds their own length from their respective dorsals, three pairs of setae on segment V, one pair of pseudostigmatals arise on or near anteromesal margin of external pseudostigmatal sockets, all ventral setae nude; posterior ventral plate with five pairs of setae, (internal poststernals lacking), all nude; tibiotarsus I with five rod-like and four, short, stout solenidia the most anterior and lateral of the latter not reaching tip of segment.

Collected from Michigan, Indiana, Georgia, from rotten stump in wet wood.

12   Gnathosoma distinctly wider than long, spherical, flattened, expanded anteriorly and laterally with a diminution and ventral displacement of the mouth-parts; chelicerae large, distinctly bladelike ....
.... Genus *Peponocara*

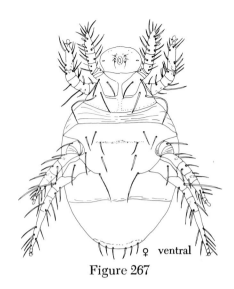

Figure 267

Figure 267   *Peponocara cathistes* Cross

With the characters of the genus. This is the only species contained in *Peponocara* collected from a polyphagine cockroach, *Arenivaga apacha* from Texas.

13   A pair of enormous rectangulate setae basally on tarsi II and III; gnathosoma long, acuminate, palps elongate, sharp, fitted for piercing; body elliptical, heavily sclerotized. ....
.... Genus *Glyphidomastax*

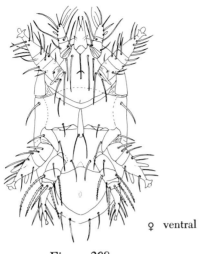

♀ ventral

Figure 268

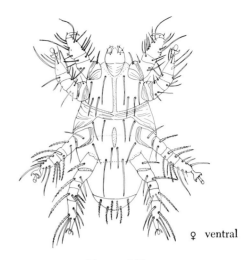

♀ ventral

Figure 269

Figure 268 *Glyphidomastax rettenmeyeri* Cross

With the characters of the genus. This species is the only one contained within the genus *Glyphidomastax*. It was erected to accommodate a new mite phoretic upon two species of North American army ants, *Neivamyrmex opacithorax* and *N. nigrescens*.

14 The longitudinal distance betwteen internal and external presternal distinctly less than that between external presternal and first axillary; leg I shorter and thinner than leg II, claw I may be enlarged .... Genus *Parapygmephorus*

Figure 269 *Parapygmephorus halictinis* Cross

This species along with several undescribed species were reported by Cross (1965) as being phoretic upon bees. One of the undescribed forms was collected from woodland leaf litter. *P. halictinis* was collected from the halictine bee, *Agapostemon virescens* in huge masses on the anterior face of the first metasomal tergum.

Collected from Ohio and Indiana.

15 Leg I at least as wide as leg II, longitudinal distance between internal presternal and external presternal always equal to or greater than transverse distance between external presternal and first axillary ....
.... Genus *Pseudopygmephorus*

Many of the species contained in this genus were placed here by Cross (1965) from the genus *Pygmephorus*. Species known to be found in the United States are normally associated with soil and small mammals.

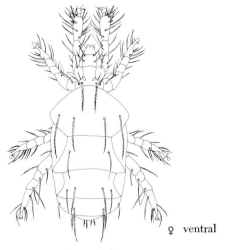

Figure 270

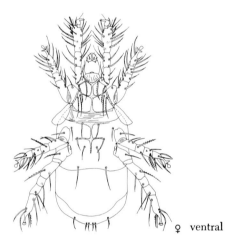

Figure 271

Figure 270 and 271   *Pseudopygmephorus tarsalis* (Hirst)

Cross (1965) recorded the possibility that he had collected *P. tarsalis* within the United States.

16   Coxa IV broadly constricted, about two-thirds of distance from apex, base rounded, bulbose; if constriction not distinct, then legs long and thin; trochanter

IV at least two and one-half times as long as wide .... Genus *Neopygmephorus*

Figure 272

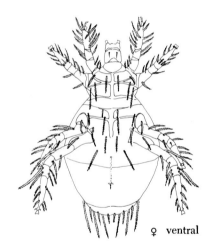

Figure 273

Figure 272 and 273   *Neopygmephorus sp.*

Collected from Kansas in rodent cache under log.

Family Cheyletidae

The family Cheyletidae has been recently reviewed by Summers and Price (1970). In this work they included approximately 50 genera and 186 species. The members of this family may be associated with birds and mammals in which they act as free-living predators on small invertebrates. They also feed on many other microarthropods such as acarid mites, collembolans and scale insects. Cheyletids are abundant in granaries, warehouses, barns, stables, where acarid populations are in high numbers. They will be found in leaf litter, topsoil, tree bark, foliage of trees where they feed on scale insects and plant sucking mites. They are considered to be important in biological control of some economic plant infesting pests.

1   One pair of eyes present; dorsal plating represented by one or more well-defined sclerites, dorsal plating of hysterosoma consists of one pair of sclerites partly covering metapodosoma; dorsal setae lanceolate or narrow fans, one seta in each of the large paired plates covering metapodosoma .... **Genus *Cheletomimus***

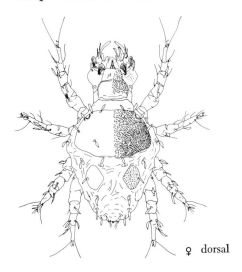

♀ dorsal

Figure 274

Figure 274 *Cheletomimus duosetosus* Muma

Female with one pair of median setae on propodosomal plate; two setae on genu IV, three pairs of median propodosomal setae, microtúbercles on tegmen circular, close together; six to eight teeth on palpal claw.

Collected from citrus litter, sand and pine litter in Florida.

2   Dorsal setae uniformly rounded or clamshell in outline; seven setae on each large paired plate covering metapodosoma .... **Genus *Oudemansicheyla***

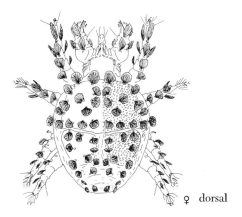

♀ dorsal

Figure 275

Figure 275 *Oudemansicheyla denmarki* (Yunker)

Small mites; palp claw partly enveloped by two leaf-like setae of palp tibia, toothed on entire margin with 15 teeth; eyes present; dorsum with tubercles some forming irregularly swollen rugae.

Collected from leaf litter of citrus and palmetto in Florida.

3   Pedicel of tarsus one normal, about as long as any setae arising on it; claw on

tarsus one smaller than those on tarsus two but not difficult to discern; mesal paraterminal sensillum of tarsus one setiform not appreciably overreaching tips of claws .... **Genus Hemicheyletia**

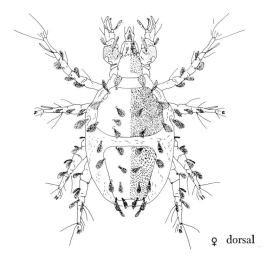

Figure 276

Figure 276 *Hemicheyletia rostella* Summers and Price

Female with three pairs of dorsomedian setae on hysterosomal plate; palp claw with 11 or 12 teeth; hysterosomal plate narrower than propodosomal plate, separated from it by a wide band of striae, this striae covers most of hysterosoma; dorsal setae, 14 to 16 pairs.

Collected from leaf mold and pine leaf mold in California.

4   Pedicel of tarsus I stubby, much shorter than setae arising on it; claws on tarsus are exceptionally small, barely perceptible; mesal paraterminal sensillum of tarsus I solenidiform, at least half as long as addorsal seta of corresponding side .... **Genus *Paracheyletia***

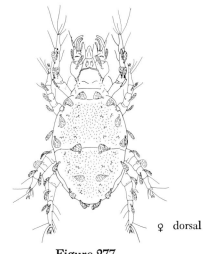

♀ dorsal

Figure 277

Figure 277 *Paracheyletia pyriformis* (Banks)

Palp claw bears 12 or 13 teeth; dorsal seta on palp femur and genu robust, lanceolate, with rows of barbs closely appressed; body plates cover entire dorsum; propodosomal and hysterosomal plates nearly equal in width, contiguous or overlapping at humeral sulcus; eyes large; dorsal setae number, 23 pairs.

Collected from lilac and poplar leaves in Idaho.

5   **Dorsal plating coarsely reticulate; legs, short; two principal body plates; genital and anal covers lie close together to form a broad, ovoid assembly ....**
.... **Genus *Ker***

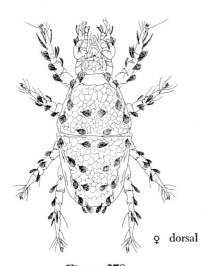

♀ dorsal

Figure 278

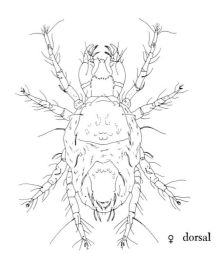

♀ dorsal

Figure 279

Figure 278 *Ker palmatus* Muma

Dorsal plating with well developed porosities or microtubercles; dorsal setae fan-shaped; tibia I with five setae, solenidion on tibia I inflated and long enough to project beyond the tibiotarsal flexure by at least half its length.

Collected from leaf litter in Florida.

6   Setae on margins of dorsal plates acicular, fusiform or narrowly spatulate, conservatively barbed; fan-shaped anal setae absent; dorsomedian setae, when present, few in number, tiny and simple in structure .... **Genus *Cheyletus***

Figure 279 *Cheyletus eduditus* (Schrank)

Female femur IV with two setae; palp claw with two similar basal teeth; tegmen ornamented with broken striae and anastomosing trabeculae between origins of muscle bundles; propodosomal plate covering body outline; hysterosomal plate trapezoidal, front margin arched, incompletely covering opisthosoma.

Collected from floor of dairy barn, eucalyptus bark, moss, haybarn litter in California.

7   Setae on margins of dorsal plates broadly spatulate or fan-shaped; dorsomedian setae numerous, conspicuous, greatly modified .... **Genus *Eucheyletia***

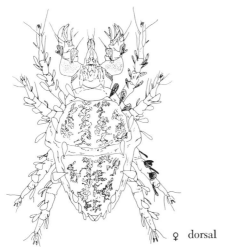

♀ dorsal

Figure 280

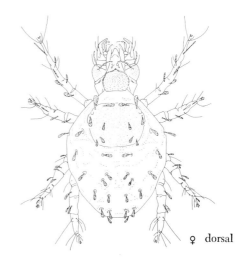

♀ dorsal

Figure 281

Figure 280   *Eucheyletia bishoppi* Baker

Female with each median dorsal plate with seven pair of cloud-like median setae; large individuals; palp with basal teeth; propodosomal plate concave, heavily sclerotized, sharply outlined; posterior pair of anal setae spatulate.

Collected from soil samples, leaf mold, *Neotoma* nest, and topsoil beneath Bishop pine in California.

8   Dorsolateral setae on hysterosoma rod-like, strap-like, spatulate, or fan-like; one to six pairs of median hysterosomal setae; median dorsal plate absent on hysterosoma .... **Genus *Cheletacarus***

Figure 281   *Cheletacarus gryphus* Summers and Price

Female palp claw with 12 to 14 teeth on mesal surface; propodosomal shield weakly sclerotized, with linear strands of tiny, equispaced rod-like elevations (dotted striae).

Collected from bark of almond tree in California (only a single specimen has been collected).

9   Idiosoma ovoid or pyriform; coxae III and IV not appreciably separated from coxae I and II; no major plate on hysterosoma; protegmen and rostrum almost coextensive, the latter barely protruding beneath the former ....
.... **Genus *Cheyletonella***

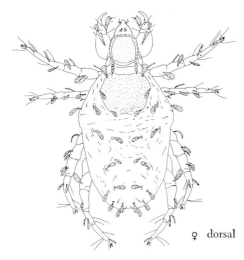

Figure 282

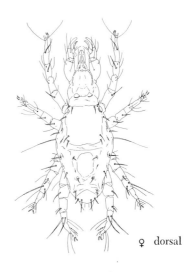

♀ dorsal

Figure 283

Figure 282 *Cheyletonella vespertilionis* Womersley

Female fourth pair or dorsolateral propodosomal setae approximately aligned in a straight crossrow with the single pair of dorsomedians on this part of the body; two of three pairs of setae on the anal swelling forked, close to apex.

    Collected from bat guano and *Pinus ariatata* soil in California. *Eptesicus fuscus* from Indiana.

10   **Dorsal body setae smooth, humeral setae ultralong; suranal plate covers tip of opisthosoma** ...... **Genus *Paracaropsis***

Figure 283   *Paracaropsis travisi* (Baker)

Female palp claw bears eight to nine teeth; eyes protuberant; hysterosoma bears two median plates, a small anterior plate in an area bounded by first and second pair of dorsomedian setae, and a small suranal plate on which are born the first three pairs of dorsal setae; dorsal body setae total 14 pairs, short, bluntly pointed not obviously barbed, many set in individual, trivial platelets.

    Collected from *Scelopores woodi* a lizard from Georgia.

11   **Palp tarsus two comb-like setae; dorsal setae long, rod-like, coarsely barbed ....**
.... **Genus *Nodele***

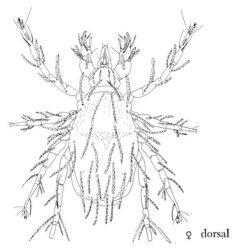

♀ dorsal

Figure 284

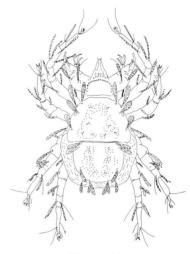

Figure 285

Figure 284 *Nodele calamondin* Muma

Female guard setae smooth; palpal tibia carries claw on robust pedestal having no blade-like flange on its dorsomesal face; dorsal plating very feebly sclerotized, boundaries indeterminate; three pairs of anal setae, dorsal most pair much longer than other two pairs.

Collected from leaf litter of Monterey pine in California.

12   Leg I equal to or longer than idiosoma, palp claw with seven to 12 teeth; two setae on femur IV .... Genus *Mexecheles*

Figure 285 *Mexecheles virginiensis* Baker

Palpal claw with nine to ten teeth; dorsum of rostrum with few tubercles and reticulate pattern; dorsal body setae long and broadly lanceolate; ventrally dorsolateral anterior body setae one considerably exceeds distance from its base to dorsal midline.

Collected from pine bark, associated with bark beetle, *Ips,* tree bark, associated with *Dendroctonus frontalis* (bark beetle) in Virginia, Nevada and Utah.

13   Marginal setae of propodosoma heteromorphic, those in front of eyes broad, flabellae, those behind eyes long, narrow, strap-like .... Genus *Grallacheles*

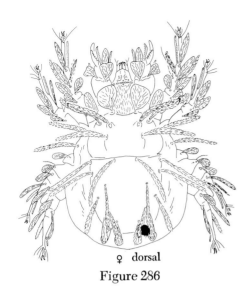

♀ dorsal

Figure 286

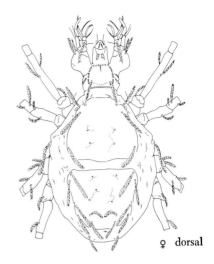

♀ dorsal

Figure 287

Figure 286   *Grallacheles bakeri* DeLeon

Palpal claw bears six conical teeth; dorsal body setae 15 pairs, on tergal plating; peritremes inconspicuous; palps and ambulatory legs with foliate setae, a few flabellate, majority fan-like to spatulate; two smooth plates covering most of dorsum, plating not obviously ornamented; eye present.

Collected from bark of citrus trees in Florida.

**14   Palp claw with one tooth, one seta on femur IV; dorsal body setae relatively long, stout, rod-like, four pairs of marginal setae on propodosomal plate; leg I longer than idiosoma ....
.... Genus *Cheletomorpha***

Figure 287   *Cheletomorpha lepidopterorum* (Shaw)

Gnathosoma projects well in front of body; basis capituli fairly long; dorsal body setae number 15 pairs; forelegs very long and slender; palp claw slender, bent near base, tip almost straight and spinelike; one pair of eyes; dorsal lateral setae stout, rodlike, blunt tipped, densely barbed.

Collected from cattle feed bins, dairy barn floors, bark of ponderosa pine in Texas, Virginia, Maryland, Kansas, and California.

15   Tarsus I with only two conspicuous terminal setae .... Genus *Cheletogenes*

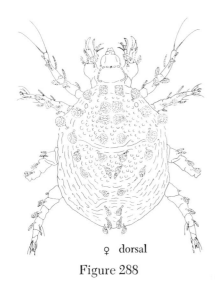

♀ dorsal

Figure 288

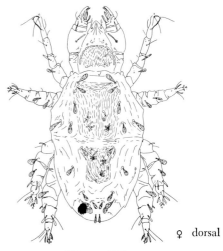

♀ dorsal

Figure 289

Figure 288 *Cheletogenes ornatus* (Canestrini and Franzago)

A small rotund species having a papillose body integument and short palplike forelegs; palp claw with 15 teeth; two dorsal plates, propodosomal plate covers area bounded by four pairs of marginal setae; hysterosomal plating not well defined; dorsal body setae number 15 pairs.

    Collected from almond twigs, pansy foliage, oak twigs, lichens, bark of peach trees, sagebrush, in California, Utah, Florida, and Louisiana.

16   **Tarsus I with at least four conspicuous terminal setae .... Genus *Prosocheyla***

Figure 289 *Prosocheyla buckneri* (Baker)

Female with first and second pairs of dorsolateral setae on margin of propodosomal plate, third and fourth pairs on independent platelets; dorsal body setae number 15 pairs; propodosomal shield indicated by thickened longitudinal striae covering area between eyes.

    Collected from citrus leaves and lemon in Florida and California.

17   **With 6 pairs of dorsolateral setae on margin of propodosomal plate; palp claw without teeth .... Genus *Eutogenes***

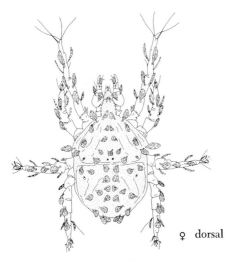

♀ dorsal

Figure 290

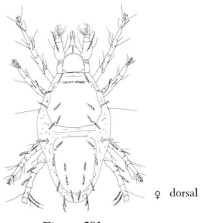

♀ dorsal

Figure 291

Figure 290 *Eutogenes foxi* Baker

Females dorsal setae appear as broad, rounded fans, each with a pronounced skirt; dorsal seta of palpal femur leaflike, not a fluted rod, palpal claw smooth, dentate; entire dorsum covered with two roughened plates; dorsal body setae numerous, 25 pairs; propodosomal plate bears six pairs of laterals and six pairs of medians.

Collected from citrus leaf litter, rose stems in Florida and Texas.

18   Humeral setae acicular, smooth, or nearly so, much longer than dorsal body setae; palpal tarsus with a sickle, one orthotox comblike setae ....
.... **Genus** *Acaropsis*

Figure 291 *Acaropsis sollers* Rohdendorf

Palpal femur two times as long as it is wide; dorsal seta of palpal femur longer than femoral segment; third pair of medioventral body setae aligned with setae on covae IV; paragenital setae twice as long as genital setae; eyes inconspicuous; hysterosomal plate truncate in front, mended behind.

Collected from sweepings from cattle pens and feed trash, swine barns in California.

19   Humeral setae relatively short and spatulate, resembling nearly the dorsal setae .... **Genus** *Acaropsella*

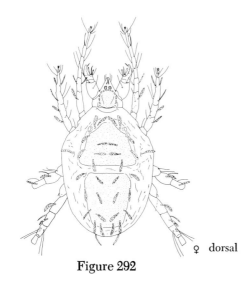

Figure 292

♀ dorsal

Figure 293

Figure 292   *Acaropsella kulagini* (Rohden-dorf)

Female with dorsal seta of palpal femur acicular, finely barbed, drawn to a slender point; several anal setae forked; palpal claw bears three to five sharply pointed teeth; dorsal plates clearly outlined, provided with fine, broken striae, transverse in direction; eyes present sharply delimited.

Collected from bark of grape vines, deep litter in hay barn in California.

20   Dorsal plates bear about 16 to 18 pairs of squamate dorsomedian setae; palpal claw with eight to eleven small teeth distributed along most of its length .... Genus *Hypopicheyla*

Figure 293   *Hypopicheyla elongata* Volgin

The anogenital covers are displaced far forward so that the genital plates lie close behind coxae IV; distance between anal papilla, the hind margin of opisthosoma exceeds greatest width of entire assembly of anogenital sclerites; dorsal hysterosomal plate tucks under rear part of body; sixth pair of dorso-laterals arise on ventral surface farther forward than the fifth pair; setae of fifth and sixth pairs do not align in a straight crossrow.

Collected from soil beneath pine and vineyard soil in California.

21   Inner sensillum, generally called the sicklelike seta of palp tarsus, distended or inflated .... Genus *Neoeucheyla*

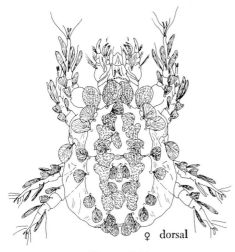

Figure 294

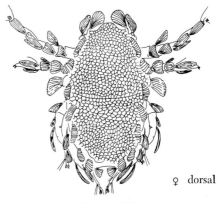

♀ dorsal

Figure 295

Figure 294 *Neoeucheyla typhosa* Summers and Price

Female dorsal setae on palpal femur forked, with five or six prongs; palpal claw gently bowed with one pointed tooth; dorsal and dorso-lateral setae on palpal femur peculiarly modified, blade of each seta forklike with five or six branched spikelets.

Collected from soil under pine tree in California.

22 **Gnathosoma foreshortened, with usual projection of rostrum from apex of basis capituli disarranged by midventral encroachment of palpal femora; no subcapitular setae; peritremes from reverse loops around paired chambers in margins of stylophore .... Genus *Cunliffella***

Figure 295 *Cunliffella whartoni* (Baker)

Female dorsomedian setae all alike, so overlapped that their separate outlines and pedicels are obscure; four teeth on palpal claw.

Collected from Wisconsin, habitat unknown.

23 **Two types of flattened dorsal body setae, dorsolaterals labellate, dorsomedians squamate, decumbent; eyes present .... Genus *Microcheyla***

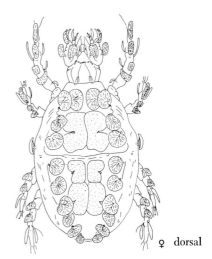

♀ dorsal

Figure 296

Figure 296 *Microcheyla parvula* Volgin

Empodia present, claws absent on all tarsi; tarsi II to IV bearing only 4-4-4 setae; palpal claw comblike, bears numerous slender teeth; eyes present; rostrum pointed; superior aboral setae set on projecting nipples.

Collected from soil and leaf trash under juniper in California.

24   **With no comblike seta on palp tarsus; anal and genital covers contiguous, in subterminal position** .... **Genus *Alliea***

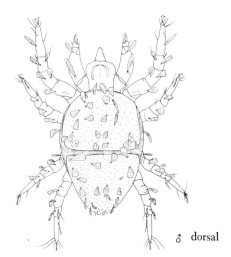

♂ dorsal

Figure 297

Figure 297 *Alliea laruei* Yunker

With the characters of the genus, only a single collection of this species from *Rattus norvegicus* in Florida.

25   **Hysterosoma without a plate, protegmen and rostrum almost coextensive, the latter barely protruding beneath the former** .... **Genus *Cheletonella***

A single species of this genus, *C. yasguri* was collected from a domestic dog, in New York.

26   **Peritremal segments not conspicuously widened in descending arms, similar in size and shape to those of transverse arms; propodosomal shield without median setae near posterior margin; no ultralong setae on opisthosoma** ....
.... **Genus *Eucheyletiella***

A species *E. johnstoni* was collected from *Ochotona princeps* from New Mexico.

27   **Coxa III distant from IV; coxa II obsolete, without sclerotized margins** ....
.... **Genus *Neocheyletiella***

Collected from *Leucosticte australis* and a robin, *Turdus migratorius*, in Colorado and Delaware.

28   **Idiosoma slender, elongate, appreciably constricted in midsection, coxae II and III separated by more than body width, body setae acicular, smooth or minutely barbed** .... **Genus *Chelacheles***

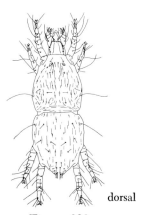

dorsal

Figure 298

Figure 298 *Chelacheles bipanus* Summers and Price

Humeral setae of propodosoma and first pair of dorsolateral setae of hysterosoma are ultralong, palpal claw pointed, shorter than tibial joint, 3-4 teeth, propodosomal shield represented by faintly tanned patch of cuticle shaped like inverted triangle, its base line wedged between eyes, dorsal body setae, 13 pairs.

Collected from willow bark and twigs in California.

29 **Idiosoma fusiform, elongate, coxae of legs III and IV displaced far to rear, so that coxa IV encroaches upon genital covers** .... **Genus** *Bak*

Figure 299 *Bak sanctaehelenae* Yunker

Each peritreme has a median descending arm comprising 3 links, claw on legs I to IV sharply bent, each with a basal swelling, palpal claw a robust crescentic spur, propodosomal plate feebly sclerotized, covers tergal area bounded by 4 pairs of propodosomal setae, dorsal setae spiniform, very short, 11 pairs.

Collected from topsoil in California.

♀ dorsal

Figure 299

Family Tenuipalpidae

Members of this family are known as the false spider mites. They feed on the leaves of plants, and are often found on the lower surface of the leaf. There are, however, several species within the family that feed on the bark of small plants, floral heads and under leaf sheaths of certain grasses. Some species are capable of causing plant galls. Their mouthparts are well adapted for feeding on plants as the cheliceral stylets are needle-like and capable of piercing the epidermis. This type of feeding habit results in the removal of chlorophyll which in turn causes the plant tissues to become silvery turning to a rust color. This is characteristic of false spider mite feeding damage.

A number of species within the tenuipalpids are of economic importance; *Brevipalpus californicus*, *B. phoenicis*, *B. oncidii* and *Tenuipalpus pacificus* attack orchids grown in greenhouses.

Members of this family may be a bright red, green or yellowish green due to their feeding habit. The gall forming species may have a relationship to other gall forming mites such as the eriophyid mites.

A worker that attempts to identify members of this family should consult the works of Pritchard and Baker (1951 and 1957).

1 **Female with four pairs of legs and three pairs of anal setae; male with four pairs of genito-anal setae; palpus with four or five segments, hysterosoma with four pairs of dorsosublateral setae ....**
.... Genus *Aegyptobia*

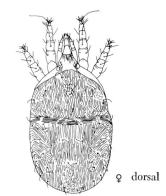

♀ dorsal

Figure 301

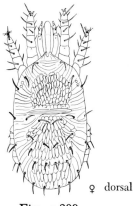

♀ dorsal

Figure 300

Figure 300   *Aegyptobia nothus* (Pritchard and Baker)

Female rostral shield deeply emarginate, caudolateral hysterosomals and distal propodosomals dissimilar; body with dorsal integument reticulate; idiosoma with dorsal setae slender.

Collected on *Juniperus* sp. from Oklahoma.

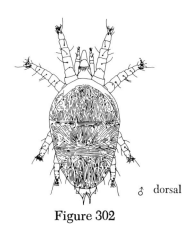

♂ dorsal

Figure 302

Figures 301 and 302   *Aegyptobia aplopappi* Baker and Tuttle

Female lacks hysterosomal pores; dorsal and marginal body setae narrowly lanceolate, nude; hysterosomal striae longitudinal except for a marginal area; rostrum slender, reaching past end of genu I; dorsal setae of femora I and II slightly lanceolate; setae of genu and tibiae I and II setiform; tarsi I and II each with a single rod-like sensory seta; claws well developed; rostral shield absent; anterior margin of propodosoma slightly indented; propodosoma striae longitudinal, with small reticulate area just behind first pair of propodosomals; setae of medium length, nude, slightly lanceolate, hysterosomal striae longitudinal except for marginal area; setae similar to those on propodosoma but shorter.

Collected on *Aplopappus acradenius* from Arizona.

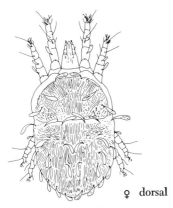

♀ dorsal

Figure 303

Figure 303 *Aegyptobia desertorum* Baker and Tuttle

Female possesses hysterosomal pores behind the second pair of dorsosublateral setae; femora, genu I and II and tibiae I and II possess a strongly clavate serrate seta; no rostral shield; striae pattern of the propodosoma longitudinal; propodosomal setae long, clavate, serrate; striation pattern and setae of the hysterosoma are similar to those of the propodosoma; rostrum long, reaching nearly to tip of genu I; dorsal setae of femora and genu I and II broadly clavate, serrate; that on tibiae I and II long, whip-like, tarsi I and II each with a long single sensory rod; claws strong; rostral shield lacking, anterior margin of propodosoma broadly rounded; striae of propodosoma longitudinal, forming long reticulations; setae long, subequal, clavate, serrate; hysterosoma with pores, striae as on propodosoma except for transverse area anterior to second pair of dorsosublateral setae, setae similar to propodosomal setae. Genital and ventral plates without striae.

Collected on *Atriplex canescens* from Arizona.

2   Rostral shield with narrow, acutely pointed lobes; female without a ventral plate; hysterosoma with two pairs of dorsosublaterals; palpus with four or five segments .... Genus *Pseudoleptus*

Collected on *Distichlis spicata*, crab grass, *Hilaria mutica, Hilaria rigida, Tridens pulchellus, Guterrezia sarothrae* from California, Kansas, Colorado and Arizona.

3   Rostral shield, when present, incised, and with broad lobes; female with ventral plate; palpal segments five in number; body broadly ovate ....
.... Genus *Pentamerismus*

♀ dorsal

Figure 304

Figure 304 *Pentamerismus erythreus* (Ewing)

Female dorsocentral hysterosomals minute, seven in number.

Collected on *Cupressus, Libocedrus, Chamecyparis, Thuja, Juniperus, Juniperus deppeana, Cupressus sempervirens,* accidental on sequoia, *Picea, Olea, Rubus,* and *Hydran-*

gea from California, Washington, Oregon, Pennsylvania, Ohio, Virginia, Washington, Kansas, Washington, D.C. and Arizona.

4   **Hysterosoma without dorsosublaterals; palpus with four segments (figs. 305, 306, 307, 308, 309, 310, 311, 312)** .... **Genus *Brevipalpus***

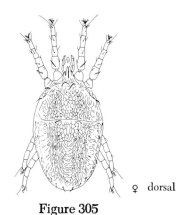

Figure 305   ♀ dorsal

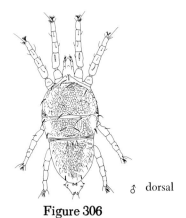

Figure 306   ♂ dorsal

Figures 305 and 306   *Brevipalpus artemesiae*

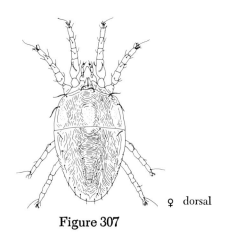

Figure 307   ♀ dorsal

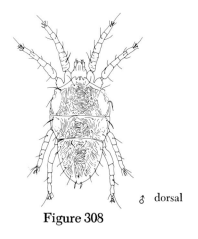

Figure 308   ♂ dorsal

Figures 307 and 308   *Brevipalpus allenrolfeae*

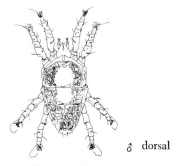

Figure 309   ♂ dorsal

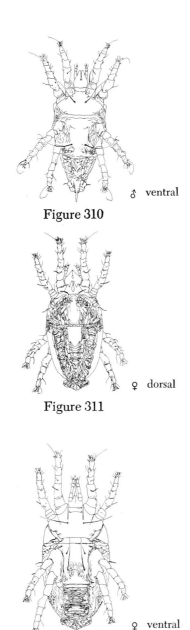

♂ ventral

Figure 310

♀ dorsal

Figure 311

♀ ventral

Figure 312

Figures 309 through 312 *Brevipalpus colum-biensis*

This is a very large genus and contains some of the important economic species including *B. californicus*, *B. phoenicis* and *B. oboratus*,

all of which are found associated with citrus. Any worker desiring to study seriously this group and determine species is referred to the works of Pritchard and Baker 1951-1958, published by the University of California Publications in Entomology, Volumes 9 and 14.

5   Podosoma very broad; opisthosoma narrow; palpus with three or less segments; invagination for stylophore lacks longitudinal ribs; rostral shield developed in deutonymphal stage ....
.... **Genus *Tenuipalpus***

♀ dorsal

Figure 313

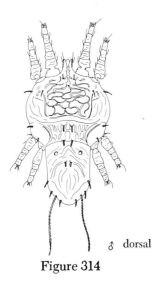

♂ dorsal

Figure 314

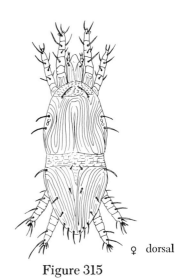

♀ dorsal

Figure 315

Figures 313 and 314 *Tenuipalpus tetrazygiae* DeLeon

Female dorsocentral hysterosomals narrowly lanceolate, third pair of dorsocentral hysterosomals widened; dorsal surface of body irregularly ridged; palpus with sensory seta on terminal segment; tarsi I and II with posterodistal sensilla.

Collected on *Tetrazygia bicolor* from Florida.

6   **Dorsal seta of femur II short; palpal setae located on dorsal region; propodosoma striated; dorsocentrals short, and simple; humeral seta serrate .... Genus *Dolichotetranychus***

Figure 315 *Dolichotetranychus floridanus* Banks

Female with short dorsal seta on femur II; two pairs of anal setae; smooth striae on genital plate; palpus with dorsal setae on third segment; propodosoma dorsally with short striae; dorsocentrals short, simple; humerals slightly longer than dorsocentrals and serrate.

Collected on *Sporobolus cryptandrus, Bouteloa gracilis* and pineapple throughout the world. Collected in the United States from Florida, Oklahoma and Utah.

Family Johnstonianidae

This family was at one time considered as a subfamily of the Trombidiidae. Newell (1957) established the family Johnstonianidae for this group. Interest has been stimulated by the possible primitive nature of the family. All three life stages, larval, nymphal, and adult are terrestrial in habit. However, as Newell (1957) pointed out, rarely are members of this family found very far from an ample supply of water and could be considered as subaquatic. The larvae of the Johnstonianidae are all parasitic, and are considered to be the "self-detaching larvae" type. Known hosts for members of this family are certain pupae of aquatic beetles found under rocks and aquatic diptera, identified as resembling the dipterious family Drosophilidae. Cold-stenothermal species, *Lassenia lasseni* Newell and *Johnstoniana latiscuta* Newell have been recorded as associated with cold mountain streams.

The adults of the subfamily Johnstonianinae have either one or two pairs of sensilla on scutum. The pregenital tubercle is absent; solenidia number two, typically clavate; supracoxal setae absent from gnathosoma and coxa I in both larva and adult. Larvae without anal sclerites; usually with a single pair of setae in intercoxal area between coxae III. Deutorostral setae absent; terminal seta of palpal tarsus not euphathidiform. Tarsi each with 2 claws.

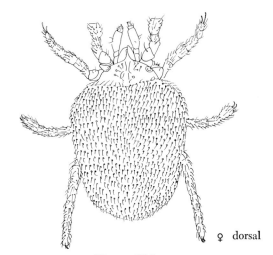

♀ dorsal

Figure 316

1a Two pair of sensilla present 2

1b One pair of sensilla present; adults with scutum greatly reduced, bearing a single pair of sensilla anterior to which is a single pair of normal setae; euphathidia confined to tarsi; larvae with two pair of setae on scutum, one pair of normal setae plus one pair of sensilla .... Genus *Centrotrombidium*

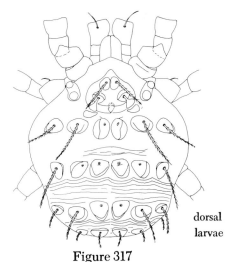

dorsal larvae

Figure 317

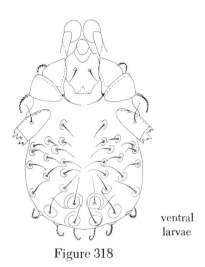

ventral
larvae

Figure 318

Figures 316, 317, and 318 *Centrotrombidium distans* Newell

Clavate solenidion two and the famulus on tarsus II widely separated, famulus borne on a protruding vesicular base; 34-41 solenidia on the dorsal surface of tibia I; ocular plates with only one distinct cornea, plus a non-refractile posterior lobe; dorsal hysterosomal setae relatively long, slender, tapering uniformly to a very fine point.

Collected from California, along alkali-encrusted shore of lake; on mud overgrown with grasses and other plants.

2a Vestigial setae absent from patella and tibia of all legs; anterior sensilla present, similar in form to posterior sensilla; anal sclerites present; eupathidia present on leg segments other than the tarsi; larvae with four pairs of setae on scutum; vestigial setae absent from patella and tibia of all legs; femora I-III of all legs completely divided into basifemur and telofemur .... Genus *Diplothrombium*

2b Vestigial setae present at least on patella and tibia 3

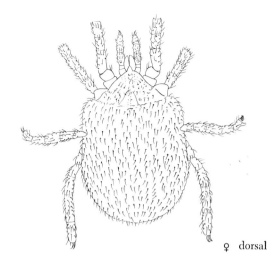

♀ dorsal

Figure 319

Figure 319 *Diplothrombium micidium* Newell

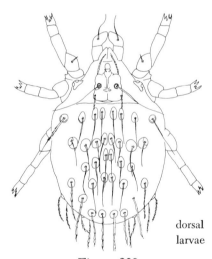

dorsal
larvae

Figure 320

Figure 320 *Diplothrombium* sp.

Female propodosomal cuticle containing about 20 setae on each side between ocular plate and scutum, each of these setae borne on a small sclerite; genital sclerites with 10 and 12 setae each in a single row; paragenital sclerites with 19 and 22 setae each; male genital sclerites with 16-18 setae, paragenital sclerites with 23-26 setae.

Collected from California under rocks and sticks along a stream in a cascade meadow.

3a **Vestigial setae present on patelli I and II, and tibia I; absent on tibia II; solenidia one and two of tarsus I differing in size only, not in structure; anterior sensilla present, considerably modified, less than one-half as long as posterior sensilla; pregenital tubercle present .... Genus *Lassenia***

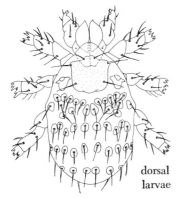

δ dorsal

Figure 321

dorsal larvae

Figure 322

Figures 321 and 322 *Lassenia lasseni* Newell

Dorsal body setae numbering more than 20, stiff, smooth, straight, each arising from a small sclerite; terminal seta of palpal tibia bifid; genital sclerites forming hemispherical protuberance, each bearing about 24 smooth, slender setae; paragenital sclerites slender, crescentic, each bearing seven setae; three pair of genital acetabula; anal sclerites well developed, slender, each having from zero to five smooth setae.

Collected from California in a small cascade stream, elevation approximately 7,250 feet. Larvae parasitic upon small species of Diptera (resembling Drosophilidae).

3b **Vestigial setae present on patella I and II, absent from tibia; anterior wall of trochanter with no trace of a fenestra; anus surrounded only by membranous cuticle containing setae, sclerites absent; eupathidia present on leg segments other than tarsi .... Genus *Johnstoniana***

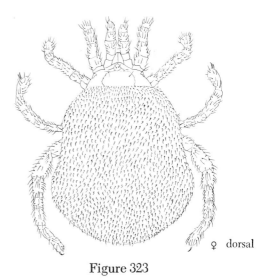

♀ dorsal

Figure 323

Figure 323 *Johnstoniana latiscuta* Newell

Vestigial setae fairly long and slender, about half the length of surrounding solenidia III, confined to patella I and II, absent from tibia; tibiae with numerous solenidia; scutum with 2 pairs of slender, smooth sensilla, plate divided into 2 portions, a broad extensive posterior portion bearing the posterior sensilla, and lateral and posterior to these six to eight setae on either side, anterior portion set off by sharp declivity and bears only the anterior pair of sensilla, in front of which is a rounded knob produced ventrally into a sharp spine.

Collected from California under rocks on edge of stream.

Remarks—This is the only species of this genus recorded from the United States and to date no males have ever been collected for the genus *Johnstoniana*.

Family Trombiculidae

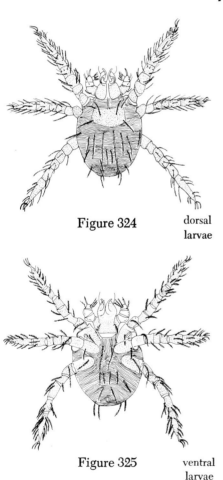

Figure 324   dorsal larvae

Figure 325   ventral larvae

Figures 324 and 325 *Trombicula (E.) alfreddugesi*

The larval members of this family are the well-known chigger that causes what the professional refers to as trombidiosis, but most people know of their effects as those small red itch welts caused by the feeding habit of larval chiggers. The most common chigger attacking humans in the United States is *Trombicula (Eutrombicula) alfreddugesi* (Oudemans) (Fig. 324, 325) and in many sections of the United States it is called the red bug. Larval trombiculids exist in a variety of different habitats throughout the United States. Baker et al. (1956) in their discussion regarding *T. alfreddugesi*, pointed out the most likely habitat a person may encounter the chigger. It was stressed that they were abundant in transiate areas between forest and grassland, along margins of swamps, blackberry patches and brush thickets. In the midwest, such as Kansas, they are widespread throughout most habitats. In many states chiggers are encountered on lawns and are quite troublesome during the period of larval development. In the south larval chiggers may be encountered throughout the entire year; however, in other sections it may be only for a period of

2 months. It is known that temperature is a limiting factor.

This family is large in the number of described species with most descriptions based on the larval stage. Few descriptions of species treat the adult form. The nymphal stage of several species has been studied by Crossley (1960) this being the only extensive work dealing with nymphal forms in the United States. Since the nymph differs from the adult in only 3 major ways: its smaller size, fewer setae; and incomplete development of genitalia; Crossley's (1960) key would aid a worker attempting to identify adult chiggers collected in the United States.

Most of the recent work involving chiggers is conducted at the National Institute of Allergy and Infectious Diseases, Rocky Mountain Laboratory, Hamilton, Montana. There is available a key to the larval chiggers of North America constructed by Brennan and Jones (1959) that a beginner would need to consult in attempting to place specimens collected in North America.

The adults of this family unlike the larva, are not parasitic on animals, but are free-living predators. They are fairly large mites usually covered with many setae giving the appearance of being hairy. They live in the soil of grasslands and woodlands, mammal nests or burrows, decaying wood crevices in rocky outcrops, bat caves, lawns, and marsh areas.

The greatest contribution to the knowledge of this group of mites would be in the collection and study of the adults and establishing the relationship between the two families, Trombidiidae and Trombiculidae.

Family Stigmaeidae

This family belongs to a group that is recognized as belonging to the superfamily Raphignathoidea. It is the largest family within the superfamily comprising about 60% of all the species presently recorded within the Raphignathoidea.

Members of the family Stigmaeidae are characterized by the presence of dorsal plates arranged into patterns. It is by these dorsal plate patterns that an observer is able to make distinctions between the different genera within the family. There are no obvious stigmata or peritremes within the members of this family, the principal internal trunks of the tracheal system are to be found between the basal sections of the chelicerae. There is a thumb-claw complex consisting of a prominent claw associated sometimes with a small accessory claw.

There is a diversity as to where members of this family are found. The genus *Ledermuelleria* may be associated with sandflies, in the soil or with different types of vegetation. Members of the genus *Stigmaeus* are commonly recovered from soil samples, whereas species of the genus *Mediolata* are found from coniferous plant, or associated with lichens. The majority of the members of the family are predaceous as exhibited by the genera *Zetzellia, Agistemus,* and some members of *Stigmaeus.* Workers attempting to identify members of this family should consult the works of Summers and Price (1961), Summers (1962-66).

1   Large hysterosomal plate overlaps sidewalls of body almost to pleuroventral

line usually with emphatic sculpturing; cheliceral bases robust, not joined together .... Genus *Ledermuelleria*

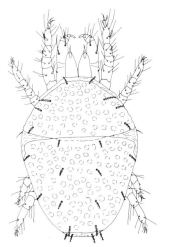

Figure 326   ♀ dorsal

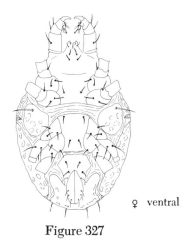

♀ ventral

Figure 327

Figures 326 and 327 *Ledermuelleria pectinata* (Ewing)

Four setae on femur II, dorsal setae bushy or burrlike, dorsal plates beset with closely spaced dimples; dorsal setae short, subequal on propodosoma, straight or slightly curved; genital plate of female widest at anterior

third, narrowed behind to width of anogenital covers; three pairs of genital setae.

Collected from California, Nevada, Utah and South Dakota from soil and litter, grassland soil, under pines, juniper, oak.

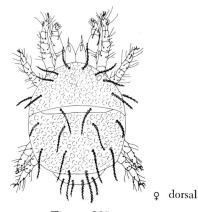

♀ dorsal

Figure 328

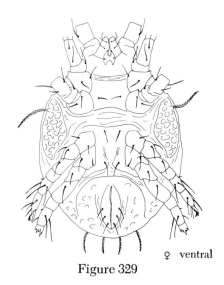

♀ ventral

Figure 329

Figures 328 and 329 *Ledermuelleria segnis* (Koch)

Dorsal setae flattened and bilaterally spinulate; one pair genital setae; postocular setae two-thirds as long as preoculars, length of each intercalary setae exceeds distance between them; dorsal plates with uniform pattern of indented dimples, these circular to oval in outline; one pair of genital setae widely spaced, three pairs of setae on anogenital covers confined to posterior half of covers.

Collected from California, Utah, from many different types of soil samples.

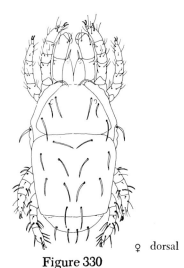

♀ dorsal

Figure 330

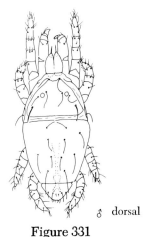

♂ dorsal

Figure 331

Figures 330 and 331 *Ledermuelleria modiola* Summers and Price

Dorsal setae flattened and bilaterally spinulate; one pair genital setae, postocular setae approximately one half as long as preocular setae, length of each seta of intercalary pair "li" less than distance between them; body small, slender; dorsal plates thinly sclerotized, coarsely dimpled.

Collected from California from grassland soil.

2   Dorsal hysterosomal plate divided; two pairs of subcapitular setae; palptarsus scarcely longer than strong tibial claw, terminal sensillum on palptarsus an obvious trident with three substantial prongs; no postocular lobes; hysterosoma in front of suranal plate covered by several plates, none with more than five pairs of setae ....
.... **Genus** *Ledermuelleriopsis*

♀ dorsal

Figure 332

Figure 332 *Ledermuelleriopsis plumosa* Willmann

Dorsal setae bushy, very short; depressions in dorsal plates sufficiently separated to appear as rounded or oval dimples; sternal plates comprise an integral prosternum and an integral metasternum, prosternum occupies venter to base of gnathosoma; genital plate covers opisthosoma almost to metasternal plate.

Collected from California from juniper leaf mold. This is the only species of this genus recorded from the United States and is one of only two described species.

3　　A stout 3-branched sensillum on the apex of palptarsus; median hysterosomal plate bears the first two or three pairs of dorsocentral hysterosomal setae and is scarcely wider than the area bounded by the origins of these setae; the dorsal eupathids on tarsis I tend to loop backward (figs. 333, 334, 335, 336) ....
.... **Genus _Stigmaeus_**

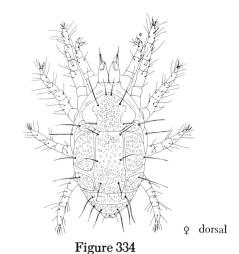

♀ dorsal

Figure 334

Figure 334　_Stigmaeus clitellus_

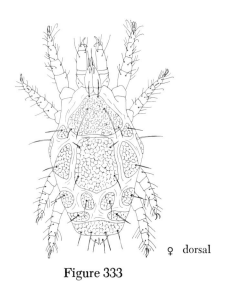

♀ dorsal

Figure 333

Figure 333　_Stigmaeus antrodes_

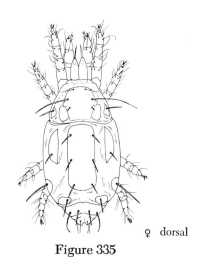

♀ dorsal

Figure 335

Figure 335　_Stigmaeus cutrichus_

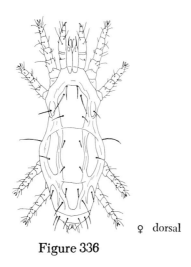

♀ dorsal

Figure 336

♀ dorsal

Figure 337

Figure 336 *Stigmaeus gracilimus*

Members of this genus are distinguished by the structure, number and disposition of plates and setae. A key to the species utilizes the number of setae which are labeled by letters of the alphabet. Summers (1962) treats thirty-one of the approximately forty species currently recorded for *Stigmaeus* in a species key.

4   Three pairs of dorsal setae on propodosoma (humeral setae excluded), setae "lm" and "c" on one median plate or on a pair of plates but never separated from each other .... **Genus Zetzellia**

Figure 337 *Zetzellia mali* (Ewing)

Two pairs of paragenital setae; median plate variously invaded by striae, with two pairs of paragenital setae, pair "a" set on isolated platelet quite remote from plate proper, setae "b" borne on areas partly detached from plate; tibia IV with four setae, no seta on genu II; median plate reticulate, not divided longitudinally.

Collected from Oregon. It is considered to be beneficial in feeding on other mites and scale insects that attack pome and stone fruit trees.

5   Intercalary plates present as a pair of small sclerites bearing only setae "li"; setae "la" on a large median plate, not isolated on separate paired platelets; dorsal plating weakly developed, compact, ovoid in shape; legs moderately long .... **Genus Agistemus**

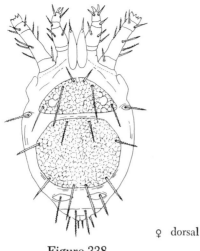

♀ dorsal

Figure 338

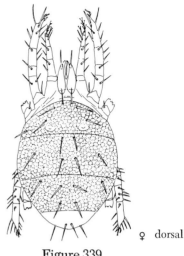

♀ dorsal

Figure 339

Figure 338 *Agistemus fleschneri* Summers

One pair of paragenital setae; main dorsal plates conspicuously reticulate, with coarse polygonal cells, reticulum on median plate with 12-14 cells in median longitudinal rows, main dorsal plates extensively ornamented; dorsal seta of medium size, regularly barbed, set on tubercles.

    Collected from California, Kansas, Nebraska, Wisconsin, Missouri, Illinois, Kentucky, Ohio, West Virginia, North Carolina, from avocado, on chaparral, willow leaf, leaf debris, log debris, apple foliage, moss, apple fruit, mulberry, sassafras leaves, poison ivy.

6   One pair of subcapitular setae; palptarsus much longer than small tibial claw, terminal sensillum of palptarsus minute, lanceolate or with 3 barely identifiable prongs; postocular bodies prominent ....
.... **Genus** *Mediolata*

Figure 339 *Mediolata pini* Canestrini

Plates extensively covering dorsal idiosoma; tibia IV with five setae; tarsal solenidion IV minute, no more than half as long as tarsal solenidion III; dorsal seta "c" longer than distance "c" to "li"; palpus long, when projected forward extending almost to distal portion of tarsus I.

    Collected from California, Colorado, from pine duff, pine leaf mold, *Pinus contorta*, rotting logs, soil litter, and soil.

7   **Propodosomal and hysterosomal plates separated by humeral suture; separate humeral plates present; large hysterosomal plate covers margins only far enough to bear lateral setae "la," "lm" and "li" .... Genus** *Cheylostigmaeus*

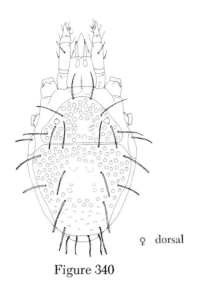

Figure 340

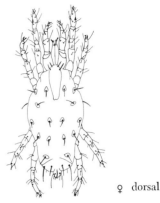

♀ dorsal

Figure 341

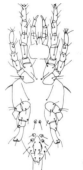

♀ ventral

Figure 342

Figure 340 *Cheylostigmaeus californica* Summers and Ehara

Male—rostral lamella a simple lobe, outer margin straight, entire anterior angle acutely rounded; uniform appendages extremely slender, finely pointed; forcipiform appendages end in incurved hooks; bulb a compact annulus closely appressed to middle of aedeagus; small species. Female with dorsum ornamented, ovoid dimples uniformly distributed on both large dorsal plates.

Collected from California from rotting logs.

8 Dorsal hysterosomal seta (except suranals) originate on very small, separate platelets; tip of palptarsus bears three or four eupathid-like setae; empodial shaft extends beyond ends of stubby claws before giving rise to pairs of raylets .... Genus *Apostigmaeus*

Figures 341 and 342 *Apostigmaeus pacificus* Summers

Thirteen pairs of plumose dorsal setae; three setae on femur IV; all four paragenital setae about equal in length; femur IV with three setae; basal pieces of chelicerae slender, fusiform, fixed digits membranous; idiosoma covered with striated integument; eyes absent.

Collected from Hawaii on *Oryza staiva, Avena stavia, Polianthus tuberosa.*

9 Dorsal setae of hysterosoma (except suranals) not set on platelets; palptarsus bears two apical setae, one of which is straight, stout and pointed; empodial shaft branches into raylets before it ex-

tends beyond ends of claws ....
.... **Genus** *Eryngiopus*

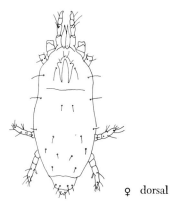

♀ dorsal

Figure 343

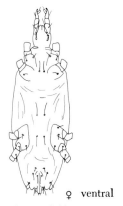

♀ ventral

Figure 344

Figures 343 and 344   *Eryngiopus gracilis*
Summers

Genu II without setae; tibia IV with four
setae; trochanter IV with one seta; three pair
of paragenital setae; propodosomal plating
represented by two narrow strips of nude integument between protruding eyes, strips
joined together in front, posterior arms of
sclerotized strips attenuated, extend almost to
humeral sulcus.

Collected from California on willow

bark. Only females of this species have been
collected.

10   Chelicerae completely joined together
close to their basal ends to form U-
shaped unit on which are located two
short straight peritremes; several pairs
of ultralong dorsal setae; each true
claw with two pair of very short, capi-
tate teneut hairs (figs. 345, 346) ....
.... **Genus** *Barbutia*

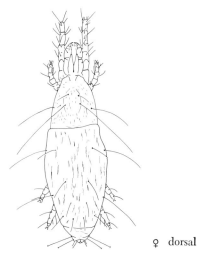

♀ dorsal

Figure 345

♀ ventral

Figure 346

Figures 345 and 346 *Barbutia anguineus*

Collected from California from mulch in a shrub thicket.

Family Tydeidae

The Tydeidae are mostly predaceous mites, associated with plants, mosses and lichens, on trees, in soils and in stored foods. According to Baker (1965), they are known to feed on scale insect eggs and plant feeding mites. There has been reports that some of the members are plant feeders. Members of this family have a worldwide distribution. In certain parts of the world species of this family have been known to attack man and his domestic animals. In stored food products it probably feeds on other mites or insects found infesting the product.

Baker (1965) stated that it would be necessary to re-collect at type localities most European species because of the damaged types in the Oudemen and Berlese collections. This creates the problem of associating species because of their wide geographical distribution. The family as treated here, follows the work of Baker (1965).

1 **Dorsal striae longitudinal on propodosoma behind sensory setae, anal opening not dorsal, striae transverse not forming reticulate pattern** .... **Genus *Tydeus***

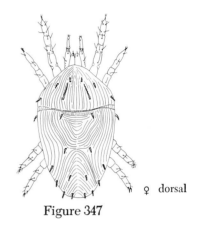

♀ dorsal

Figure 347

Figure 347 *Tydeus tuttlei* Baker

Empodia hooked; dorsal setae very short, broad, serrate; gnathosoma visible from above; chelae short; palpus with terminal segment long and slender, distal seta broadened at tip; propodosomal striae longitudinal, with small, sharp lobes, striae not reaching posteriorly to first pair of dorsocentral hysterosomal setae; sensory setae of propodosoma long, whip-like, smooth; five pairs of genital, four pairs of paragenital, one pair of anal, and three pairs of ventral setae.

Collected from Arizona on Bermuda grass.

2 **Dorsal striae forming reticulate pattern in whole or in part, L_2 setae in normal lateral position; dorsal hysterosomal setae arranged in four and one-half rows; L_5 setae are absent; six pairs of genital setae, four pairs of paragenital**

setae, one pair of anal setae and three pairs of ventral setae .... Genus *Lorryia*

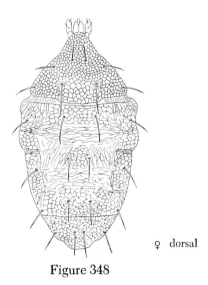

♀ dorsal

Figure 348

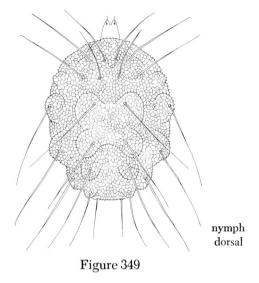

nymph
dorsal

Figure 349

Figure 348 *Lorryia bedfordensis* Evens

Reticulate pattern broken by transverse bands of elongate reticulations behind setae D_1 and D_2, three pairs of lateral rosettes formed by striation pattern; leg segments much longer and slender than other species in the genus; empodia hooked.

Collected from Washington and California on apple bark in association with *Aspidiotus lataniae* and avocado.

Figure 349 *Lorryia atyeoi* Baker

Adults unknown; nymph with dorsal body setae set on strong lobes; palpus elongate, terminal seta expanded distally; body strongly lobed, covered with reticulate pattern having small triangular lobes at each junction of striae; dorsal setae long, serrate; empodial claw present; tarsus I with small rod-like solenidion.

Collected in litter from Missouri.

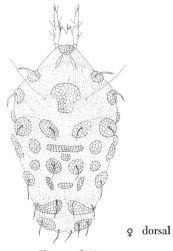

♀ dorsal

Figure 350

Figure 350 *Lorryia funki* Baker

Female body longer than broad, lobed, each lobe with reticulate pattern separated by tuberculate striae; propodosomal setae (P1) broadly lanceolate, serrate; other dorsal setae narrowly lanceolate, serrate; anal plates small, with striae, anal setae lateral from plates; empodia with small claws; solendion of tarsus I long, slender; tarsus II with short solendion.

Collected in debris of *Asyndesmus lewis* from Colorado.

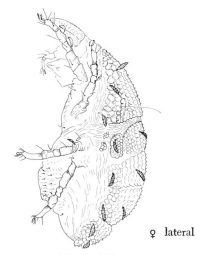

♀ lateral

Figure 351

Figure 351 *Lorryia boycei* Baker

Female body setae not broadened distally; with strong serrations; empodia with claws; six pairs of genital setae; palpal tibia with one setae; trochanter II without seta; coxa IV with one seta.

Collected under bark of camellia from Washington, D.C.

3   Dorsal striae transverse on propodosoma behind sensory setae, not forming reticulate pattern; hysterosomal setae L_2 in normal lateral position; five complete rows of hysterosomal setae ....
.... **Genus *Lasitydaeus***

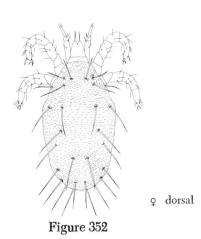

♀ dorsal

Figure 352

Figure 352 *Lasitydaeus krantzi* Baker

Female palpal setae shorter than half the length of the terminal segment; solenidia I and II of palpus as long as adjacent setae; empodia of legs strongly hooked; ventral body setae smooth and long.

Collected from California in litter.

4   With two pairs of genital setae; distal seta of palpal tarsus forked; five rows of hysterosomal setae, L_5 being present; 2 pairs of genital setae, 3 pairs of paragenital setae; hysterosomal setae L_2 in dorsal position .... **Genus *Microtydeus***

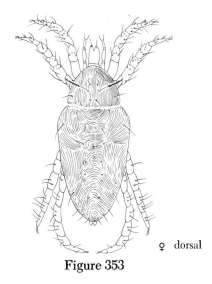

♀ dorsal

Figure 353

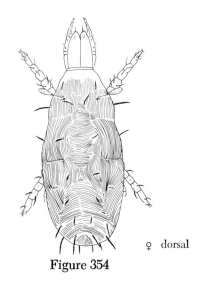

♀ dorsal

Figure 354

Figure 353   *Microtydeus beltrani* Baker

Tarsus I with 10 setae; solenidion on tarsus I and II short, broad and stalked; three dorsal setae on tarsus I strong and serrate; body setae nude; sensory setae of propodosoma pilose on distal two-thirds.

Collected from Tennessee.

5   With three pair of genital setae; striae transverse or an inverted "V"-pattern between second pair of hysterosomal setae with L_2 in dorsal position; propodosomal sensory setae club-like (fig. 354) .... **Genus *Coccotydaeolus***

Figure 354   *Coccotydaeolus krantzi*

Collected from California and South Dakota from litter.

6   With a single pair of anal setae; setae normal on femora III and IV; L_2 setae are in the dorsal position; tarsus I lacks claws and empodium; four and one-half rows of hysterosomal setae, L_5 is missing .... **Genus *Pronematus***

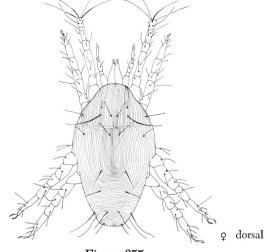

♀ dorsal

Figure 355

Figure 355 *Pronematus davisi* Baker

Female tarsus I short, possesses a small slender solenidion; posterior body setae long, slender; rostrum deeply cleft; movable chelae short; palpal distal segment elongate; propodosomal striae with minute lobes, hysterosoma striae longitudinal to setae D_2; seta D_1 and D_2 subequal, serrate; setae L_1 and L_2 shorter than D_1 and D_2; D_3 and D_4 much longer than other setae; tarsus I shorter than tibia, all distal setae longer than segment; solenidion distal, straight, slender, not as long as width of segment; solenidion of tibia I and tarsus II short.

Collected from "car window" in Indiana.

7 **Femur IV divided into basi- and Telofemur, setae L_2 dorsal, five rows of hysterosomal setae, tarsus I lacks claws and empodia ends bluntly, with four terminal setae ...... Genus *Pronematulus***

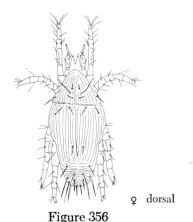

♀ dorsal

Figure 356

Figure 356 *Pronematulus vandus* Baker

Tarsus I shorter than tibia I and bearing nude setae; sensory seta of tarsus I small, does not reach distal end of segment; gnathosoma prominent, deeply cleft; propodosomal sensory setae two times longer than other setae;

strongly pilose, all body setae fine, smooth, stronger and longer posteriorly; solenidion of tarsus I not reaching tip of tarsus.

Collected from Florida on *Vanda* sp., pine needles bark, and from South Dakota on range pasture damaged by sod webworms.

8 **Striae longitudinal between second pair of hysterosomal dorsocentral setae (fig. 357) .... Genus *Paralorryia***

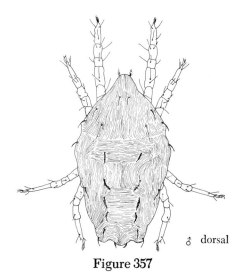

♂ dorsal

Figure 357

Figure 357 *Paralorryia sp.*

Collected from debris of *Colaptes cafer* nest in Colorado and on *Carya* bark from Pennsylvania.

9 **Claws and empodia absent on tarsi I; femur II with a distinctive strong ventral spur; five pair of lateral setae on hysterosoma; palpal setae count 5-1-2; anal setae present .... Genus *Oriola***

Collected in bark beetle tunnel from Georgia.

10 **Femur IV divided, with five lateral setae; setae L_2 on middorsal line with dorsal setae; palpal setal count, 7-2-2;**

setae D_1-D_5 and L_1 and L_2 are bifurcate .... **Genus *Meyerella***

Collected on a pine branch with *Ips* sp. from Louisiana.

Family Caeculidae

Mites of this family are commonly called rake-legged mites because of the large spines found on the first pair of legs. The first rake-legged mite collected in the United States was described by Banks (1899). Until the work of Mulaik (1945) very few members of this family had been recorded from this country. Presently, 19 species have been collected in the United States. The family has a world-wide distribution with respresentatives known from Africa, Australia, Japan and the Philippines, (Mulaik and Allred 1954). Only the single genus, *Caeculus,* is treated in this work. However, Higgins and Mulaik (1961) have established the genus *Procaeculus* for two species, *C. oregonus* and *C. brevis.*

Dorsum chitinized, large cephalothoracic plate may cover mouthparts from above, large rectangular median abdominal plate; two transverse anterior and posterior abdominal plates; legs with large spines .... **Genus *Caeculus***

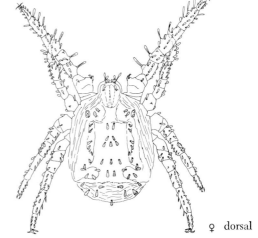

♀ dorsal

Figure 358

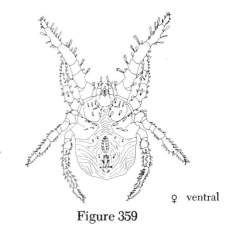

♀ ventral

Figure 359

Figures 358 and 359 *Caeculus pettiti* Nefin

Median dorsal abdominal plate with nine setae; dorsal plate of cephalothorax with four pairs of setae, first pair located at anterior margin, second pair near anterolateral margin, third and fourth pairs in posterolateral region of plate.

Collected from Virginia and North Carolina.

Family Ereynetidae

Some members of this family were formerly included in the family Speleognathidae. The two subfamilies, Ereynetinae and Speleognathinae, have contrasting habits. The Ereynetinae are predators associated with soil, mulch, or feed on the slime of snails. The species *Riccardoella limacum* (Schrank) is well known for its association with the common snail, *Helix pomata* L. in the United States. The Speleognathinae are found as inhabitants of the nasal passages of birds and mammals. The generic classification of the subfamily Ereynetinae follows that proposed by Fain (1964).

1a Dorsal sensillae represented by two pairs; genital suckers well developed, two pair present .... Subfamily Ereynetinae Oudemans 2

1b Dorsal sensillae represented by one anterior pair; genital suckers absent or vestigial; associated with the nasal cavities of birds and mammals .... Subfamily Speleognathinae Womersley 3

2a Palps three-segmented, dorsal shield and eyes absent; associated with snails and slugs .... Genus *Riccardoella* Berlese

Riccardoella limacum (Schrank) has been reported in the United States in association with the common snail, *Helix pomata* L. The life history of this mite has been studied by Turk and Phillips (1946). The whole life cycle is not spent on the snail but these mites are always in close association with the slime trail left by the snail. Both mite and snail appear to live in perfect harmony with each other.

2b Palps five-segmented; dorsal shields and eyes present or absent. .... Genus *Ereynetes* Berlese

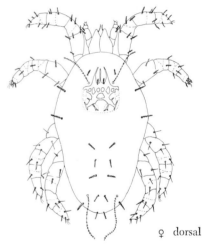

♀ dorsal

Figure 360

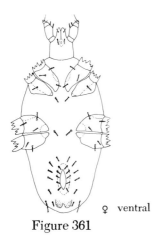

Figure 361

♀ ventral

Figures 360 and 361 *Ereynetes davisi* Hunter

Female with two pairs of anterior setae arising from above the propodosomal shield, one pair of dorsal setae (cc) in line with sensory setae; margins of propodosomal shield distinct posterior to sensillae, indistinct anterior to sensillae; dorsal setae barbed, thick; chaetotaxy of coxae as follows: 1-3, II-1, III-3, IV-2, five pairs of genital setae associated with genital opening; two pairs of distinct genital suckers; tibia I with a duplex setae arising from dorsal surface in conjunction with ereynetal organ. Male unknown.

Collected in Arizona from under bark of cottonwood, and from rotted maguey flower stalk.

3a Dorsal shield absent, eyes vestigial or absent, palps with one or two segments, polymorphic chaetotaxy with the presence of simple setae on tibiae .... **Genus *Neoboydaia* Fain**

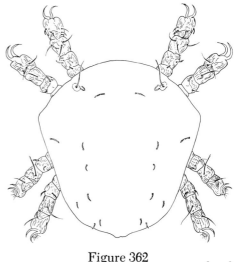

Figure 362

♀ dorsal

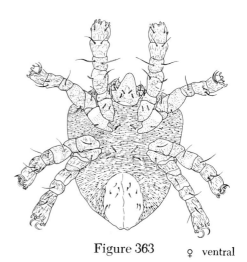

Figure 363

♀ ventral

Figures 362 and 363 *Neoboydaia colymbiformi* Clark

Tibia I with five nude, whip-like setae, single palpal segment, tarsus I with four stout recurved, nude setae ventrally, coxal setae arranged 2-1-1-1, with tightly barbelled setae on coxae I and II, simple, nude setae on coxae III and IV, eyes and scutum lacking, dorsum with a pair of nude, flagellate sensillae set in pseudostigmata, with a tiny pair of tightly barbelled presensillar setae, dorsal setae formula 2-4-2-2-4.

Collected in *Colymbus nigricollis californicus* from California.

3b Dorsal shield present 4

4a Dorsal propodosomal shield pattern present .... **Genus *Speleognathopsis***

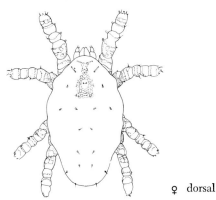

♀ dorsal

Figure 364

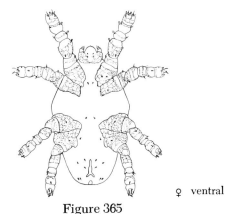

♀ ventral

Figure 365

Figures 364 and 365 *Speleognathopsis strandtmanni* Fain

Scutum reticulate, with enclosed inner cell with pair of dorsal setae, anteriorly a pair of elongate, barbelled sensillae, dorsal setae arranged 4-4-2-2-4, composed of short, ciliated,

barbelled setae, venter with three pairs of short, ciliated, barbelled sternal setae, three pairs of similar genital setae, genital plates lacking.

Collected in *Sciurus carolinensis* from Maryland. The description of *S. strandtmanni* was taken from that of *S. sciuri* Clark which was established as a synomy of *S. strandtmanni*. *S. sciuri* was collected in the United States, *S. strandtmanni* was originally collected from a squirrel, *Funisciurus carruthersi* from Africa.

4b Without a dorsal propodosomal shield pattern .... **Genus *Boydaia Womersley***

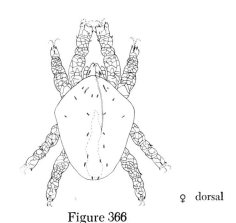

♀ dorsal

Figure 366

♀ ventral

Figure 367

Figures 366 and 367 *Boydaia sturni* (Boyd)

Larval dorsal setae consisting of 10 pairs, coxa one and trochanter one devoid of setae, tarsi II and III terminate bluntly, in two lobes between which arise two normal claws and a median bifurcated pilose pulvillus, tarsus I lacks claws and pulvillus characteristic of other legs, instead with a large double-headed claw, adult female without dorsal and ventral shields, genital aperture a median slit posterior to coxa IV, genital suckers absent, palpi three-segmented, dorsum with 11 pairs abdominal sensory setae not specialized, venter with 15 pairs of setae, genital setae arranged in two groups, two in the first group and three in the second.

Collected in *Sturnus vulgaris* from New York.

Family Demodicidae

Mites belonging to this family have long been associated with the demodectic mange of dogs. In dogs and cats this mite, when in association with species of the bacterium *Staphylococcus*, may produce a dermatitis most frequently around the eyes and muzzle of the head and on the forefeet. *Demodex folliculorum* is frequently found on humans in pores around the nose and eyelids. Many unsuccessful attempts have been made to implicate *D. folliculorum* as a cause of various skin diseases of man. This widespread species has been used in laboratory classes of both entomology and acarology to demonstrate the presence of a parasite that lives on its host and does not usually cause any disease. They may be collected from almost any person by squeezing the small pinpoint papules normally associated with the nose.

Other species of this family cause varying types of dermatitis of such animals as the pig *(D. phylloides)*, horse *(D. equi)*, cow *(D. bovis)*, sheep *(D. ovis)*, goat *(D. caprae)*, rat *(D. ratti)*, and many other animals. It should be noted that not all authors agree on species names. Some workers have treated all as belonging to *D. folliculorum*, and treated host associations as variations, such as *D. folliculorum* var. *caprae*, etc. No attempt will be made in this work to establish the taxonomic position. The work of Hirst (1919) has been followed except where factual evidence has shown otherwise.

These mites are very small and worm-like with an annulated body and eight legs which are short, stumpy, and without setae. The entire life cycle is spent on the host. They are transferred from one animal to another during contact and it is believed *D. canis* is transferred from one dog to another only during the puppy stage.

Most species of *Demodex* are very similar and species determination is a difficult task even for the specialist, let alone a beginner. Species determination for members of this family is usually based upon the host rather than the identification of the mite. The following has been adapted from Hirst (1919). It requires the use of slides with the specimens singly mounted and not the slides that are the result of the maceration-flotation technique which distorts and obliterates the morphological features required for positive species identification.

Legs short and stumpy, without setae; body without setae; palpi closely appressed to rostrum, chelicerae needle-like; female genital opening between coxae IV; male aedeagus dorsal; body worm-like with annulations .... **Genus** *Demodex*

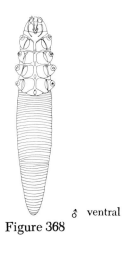

δ ventral

Figure 368

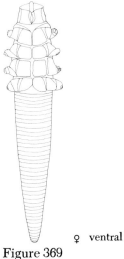

♀ ventral

Figure 369

Figures 368 and 369 *Demodex canis* Legdig

Aedeagus long, slender, curved, genital opening located above interval between the first and second legs, size of males variable. Female elongated usually more than six or seven times as long as wide, cephalothorax less than half the length of abdominal region, abdomen long, slender, tapering, posterior bluntly pointed, capitulum widest at base, two short, pointed spines on dorsal surface.

Collected from dogs throughout the United States.

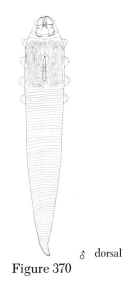

δ dorsal

Figure 370

Figure 370 *Demodex cati* Megnin

Male body five and one-half times longer than width of cephalothoracic region; abdomen one and one-half times as long as cephalothorax and capitulum; posterior end pointed, capitulum wider than long; spines on dorsal surface short, pointed, minute; genital opening placed above and in front of second pair of legs.

Collected from cats throughout the United States.

Remarks: *Demodex* found on cats do not cause the same type of dermatitis as *D. canis* produces on dogs.

Family Eupalopsellidae

The members of this family are closely allied to the family Stigmaeidae. Structural characters separating these two families are the empodium of the Stigmaeidae consisting of an erect rod-like axis on which arise directly or indirectly several pairs of empodial raylets, usually 3 pairs.

The Eupalopsellidae have at least two pairs of raylets which are sessile on a median knob of the tarsus; the erect avial rod found on members of the family Stigmaeidae, is not present. A terminal sensillum or trident is present on the palptarsus of members of the family Stigmaeidae, whereas the eupalopsellids have only a single unbranched sensillum on the apex of the palptarsus. The Eupalopsellidae, possess a pair of posterior dorsocentral setae "pm" on the propodosoma. Members of this family are found associated with citrus and leaf mold.

1a   Dorsal plate or plates ill-defined except for a pair of small indistinct areas between eyes and a narrow suranal region over tip of opisthosoma .... **Genus Saniosulus**

Figure 371   *Saniosulus nudus* Summers

Idiosoma elongated; dorsal setae extremely short; dorsomedian setae "pm" on propodosoma widely spaced; palpal claw present; a diminutive solenidum "p" on tibia; dorsum without definable plates except for a pair of small, non-striated, comma-shaped areas on propodosoma between first two pairs of setae, one small suranal plate; setae of dorsum finely denticulate, 10 pairs, very short, subequal.

Collected in Texas on orange leaf with scale insect.

1b   **Dorsal plate well-defined either covering entire idiosoma or divided into separate plates** **2**

2a   **Idiosoma incompletely covered with four unpaired dorsal plates; propodosomal, metapodosomal, opisthosomal and suranal; distal seta of palpfemur modified as a short, peculiar, brush-like blade; lateral suranal setae "le" absent** .... **Genus *Eupalopsellus***

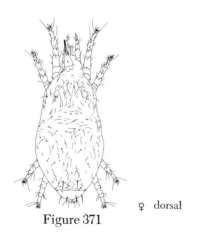

♀ dorsal

Figure 371

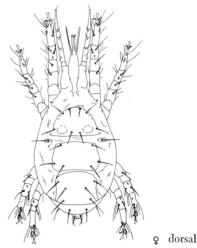

♀ dorsal

Figure 372

Figure 372 *Eupalopsellus olandicus* Sellnick

Dorsum partly covered by four median plates, each humeral setae "he" originates on its own separate humeral platelet; metapodosomal and opisthosomal plates quadrilateral, each wider than long; major plates with punctations; proximal solenidion "p" of tibiae I-IV reduced to a minute peg.

Collected from Idaho, Utah, Washington, from sagebrush.

2b **Dorsal idiosoma entirely covered by a thin skeletal sheath not clearly subdivided into discrete plates, dorsal body setae and dorsals of proximal leg segments stout, coursely denticulate, subequal, all originate on tubercles, tibial claw of palp reduced or obsolete .... Genus *Exothorhis***

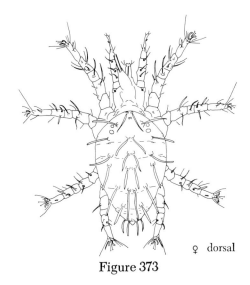

♀ dorsal

Figure 373

Figure 373 *Exothorhis caudata* Summers

Size very small; with down-curved papilla on posterior tip of opisthosoma; dorsal setae very prominent, stout, coarsely denticulate, tapered to pointed tips or bluntly rounded ends; tibia I with minute sensillum on upper distal section, this sensillum absent on other legs.

Collected in Florida from citrus leaf, grapefruit, scale infested leaf and orange leaf.

Family Pterygosomidae

Members of this family are typically parasitic on lizards and arthropods. They have attracted much attention in recent years because one of their hosts the kissing bug (*Triatoma*) is a vector of Chagas' disease. It has been established that members of the genus *Pimeliaphilus* weaken and kill bugs of the family Reduviidae and are a serious problem in maintaining laboratory colonies of these insects. The possibility of utilizing pterygosomids in the biological control of the bugs that are vectors of *Trypanosoma cruzi* Chagas has been suggested. Also the possibility that species of *Pimeliaphilus* may be able to transmit viral or rickettsial disease of their hosts is a cause for the recent interest in this group of mites. Other members of the genus *Pimeliaphilus* have been found to attract other insects such as cockroaches and scorpions.

The genera *Hirstella* and *Geckobiella* are associated with lizards and are closely related to *Pimeliaphilus* but these mites are more free-living in their habit. Several other genera within the family Pterygosomidae have been found on African lizards.

Palpus four or five segmented, with small round thumb, propodosomal shields distinct, 13 pairs of dorsal and humeral setae arranged in regular, transverse rows, paired eyes located on small plate with single seta, claws with tenent hairs, whip-like sensory setae on tibiae I, III, and IV, tarsus I with single pair of duplex sensory setae, posterior ship-like pair minute .... **Genus *Pimeliaphilus***

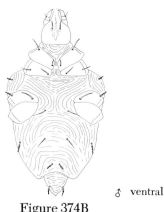

♂ ventral

Figure 374B

Figures 374A and 374B *Pimeliaphilus andersoni* Newell and Ryckman

Anterior margin of scutum concave, no trace of a division in either femur I or IV, medial seta of coxa II slender, rostral flange forming a collar-like expansion proximal to velum, wider than rostrum, ventral setae of gnathosoma smooth, slender, scutal setae 3 more widely separated, lateral seta of coxa II smooth, slender, 14 pairs of dorsal and medial setae, dorsal most setae of anal papilla aboriform, ventral setae of gnathosoma proximal to level of insertions of palpi, femur and patella of palp not fused.

Collected from Arizona and Texas on *Triatoma recurva, Triatoma gerstaeckeri.*

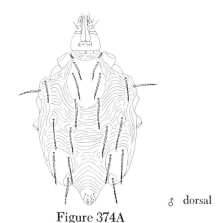

♂ dorsal

Figure 374A

Family Caligonellidae

This is a small family in regard to species numbers. Summers (1966) listed only 12 species within 6 genera. Members of this family are relatively small, reddish-colored, free-living forms. They are associated with leaf mold, tree bark, leaves, and grass litter.

1a   Peritremes forming ornate loops on its dorsal surface; dorsum of propodosoma provided with a pair of small ellipsoidal plates, one over each eye; stylophore with lobular appendages arising on its lateral-basal surface, short, thickset, basal half inflated ...... **Genus *Coptocheles***

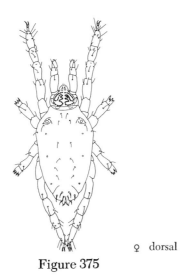

♀ dorsal

Figure 375

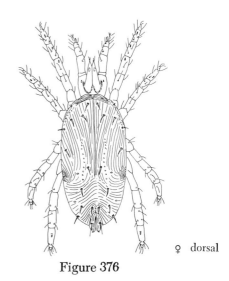

♀ dorsal

Figure 376

Figure 375 *Coptocheles bobarti* Summers and Schlinger

Peritremata emerge in middorsal line, close behind basal sclerites of movable digits to make a series of loops on inflated part of stylophore; integument covered with minutely tuberculated striae, fifth and sixth dorso-medians shorter than others, those clustered on opisthosoma stout, lanceolate; anal covers with three pairs stout, lanceolate setae.

Collected in Florida from grass sweeping and South Dakota from pastures.

1b   Peritremes not forming ornate loops
.... **2**

2a   Eyes absent; stylophore elongate, conical, tapered to bifid point in front, peritremes forming "w"-shaped structure on its dorsal surface, no obvious dorsal plates or eyes .... Genus *Neognathus*

Figure 376 *Neognathus spectabilis* (Summer and Schlinger)

Stylophore conical, twice as long as greatest width, sides slightly concave, bifid anteriorly with cleft extending backward as a groove between anchor sclerites of movable digits, peritremata approximately "w"-shaped arising on posterior fourth of stylophore, palpi stout over-reaching tip of rostrum by more than three distal joints; without minute spur or apophysis middorsally in genu of palp, claviform sensillum on dorsal aspect of femur IV.

Collected from California from manzanita leaf mold.

2b   Eyes present, stylophore not elongate; peritremes not "w"-shaped 3

3a   Stylophore without marginal lobules, peritremata arising on anterior tip of stylophore, in front of stylet bases ....
.... Genus *Caligonella*

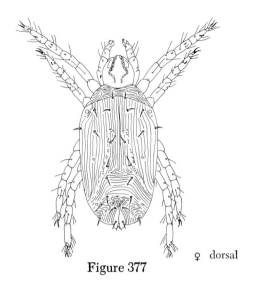

Figure 377

♀ dorsal

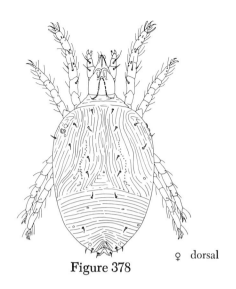

Figure 378

♀ dorsal

Figure 377 *Caligonella humilis* (Koch)

Peritremata conspicuous, confined to dorsal surface of stylophore, each appearing to communicate with tracheal system at anteriormost end of stylophore; stylophore without marginal lobules, rostrum not longer than greatest width, stubby, emphatically truncate anteriorly.

    Collected from California and South Dakota from leaf mold of juniper, pasture soil.

3b   **Stylophore without marginal lobules, peritremata arising dorsally on middle portion of stylophore, close behind stylet bases .... Genus *Molothrognathus***

Figure 378   *Molothrognathus leptostylus* Summers and Schlinger

Scapular setae not longer than other dorsal setae; slender mite, with short palpi, peritremata originate from vessels ascending between basal sclerites of movable digits, idiosoma ovoid, integument striated without identifiable dorsal plates, anal covers with two pairs of setae, legs slender, weakly developed in relation to body proportions.

    Collected from California from bark of almond trees, pine leaf mold.

Family Harpyrhynchidae

Members of this family are avian parasites that occur mostly on the contour feathers of the head, neck, and upper breast. They may be encountered on the auricular, gular, and malar feathers as well as on those of the lores and occiput. Attachment sites tend to be in areas where the host cannot preen effectively.

Mites of this family are often found by the washing of birds with water and detergent, however, because of their small size, they are commonly overlooked.

    **Legs III and IV greatly reduced; only 2 or 3 podomeres apparent; pretarsi lacking ....
.... Genus *Harpyrhynchus***

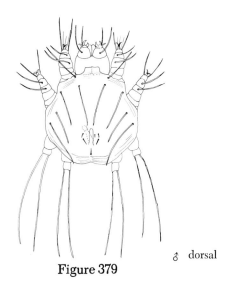

♂ dorsal

Figure 379

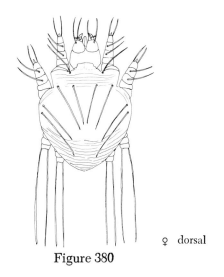

♀ dorsal

Figure 380

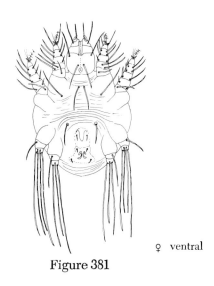

♀ ventral

Figure 381

Figures 379, 380, and 381   *Harpyrhynchus herodius* Boyd

Female terminal palpal segment with three regular setae, three elaborately developed setae of which one is a terminal claw, with 5 teeth, the other 2 are ctenoform; outer lateral and dorsal setae adjacent to stigma shorter than other dorsal setae; setae on legs III and IV short, legs I with tarsus bearing short sensory seta dorsally, feather-like pulvillus between 2 claws, leg II tarsus sensory seta narrower than same seta on tarsus I; male body much smaller than female; genital-anal region on dorsal surface; genital area provided with 3 pairs of short setae.

Collected on *Ardea h. herodis* (Great blue heron) from Massachusetts.

Harpyrhynchus butorides Boyd

Terminal segment of tarsi I and II with sclerotized rod-like ridge; palp 3-segmented, distal segment with 2 regular and 3 ctenoform setae, palpal thumb prominent, dorsum with 5 pairs long simple setae of equal length, 2 pairs of minute genito-anal setae on ventor; tarsi II with only 4 ventral setae; male with shorter legs, smaller than female, pulvilli of tarsi I and II 6-plumed, leg IV 2-segmented.

Collected on *Butorides virescens* (Eastern green heron) from Massachusetts.

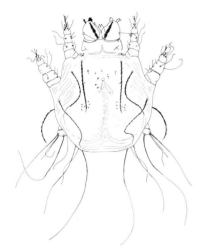

Figure 382 ♂ dorsal

Figure 383 ♂ ventral

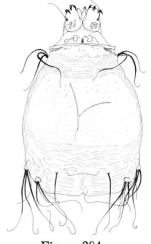

♀ dorsal

Figure 384

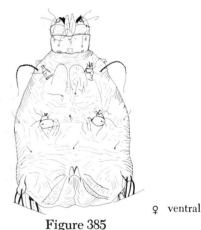

♀ ventral

Figure 385

Figures 382, 383, 384, and 385 *Harpyrhynchus novoplumaris* Moss, Oliver and Nelson

Female palpal setae (g_1) thickened, distinctly shorter than other 2 palpal setae (g_{2-3}), lobes at bases of legs I and II well developed; legs I and II with 5 segments composed of claws and bifid, pectinate empodium, legs III reduced to 2 segments bearing 4 apical, whiplike setae; single pair of subterminal posterior idiosomal setae; male also with modified palpal seta (g_1), idiosoma with large dorsal shield, broadened posterolaterally, incised posteriorly, genital opening located in dorsal shield at level of leg II, aedeagus tubular, whip-like, genital opening an inverted "V" with 3 pairs of minute setae.

Collected on *Certhia familiaris* (Brown creeper), *Parus bicolor* (Tufted titmouse), *Campylorhynchus brunneicapillus* (Cactus wren), *Richmondena cardinalis* (Cardinal), *Pipilo fuscus* (Brown towhee), *Amphispiza bilineata* (Blackthroated sparrow), *Spizella passerina* (Chipping sparrow) from Kansas,

California, Missouri, Arizona, Maryland, Nebraska, New Mexico, and Wyoming.

Two other species have been collected in the United States, *H. brevis* Ewing and *H. longipilus* Banks. Their descriptions are too general to allow a worker to separate them from existing species described in the genus *Harpyrhynchus*. *H. brevis* was collected on *Coccothraustes vespertina* and *H. longipilus,* on *Loxia* sp.

Family Neophyllobidae

Members of this family are called stilt-legged mites and were considered a part of the families Caligonellidae, Stigmaeidae and Tetranychidae. They feed on smaller mites and insects especially scale insects. They have a wide range of distribution having been collected from Virginia to California with specimens recorded from such midwestern states as Texas and Colorado. Only the single genus, *Neophyllobius*, is recorded from the United States. The work of McGregor (1950) treats species described in the United States.

Legs long, exceeding the body in both width and length; leg segments with few hairs, mostly arising from tubercles; rostrum short, at times hidden; palpi short, slender, five-segmented; dorsal body setae peg-like, lanceolate, or clavate often born on tubercles .... Genus *Neophyllobius*

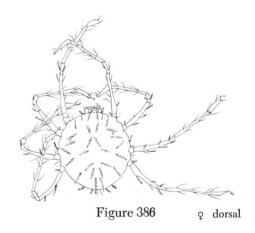

Figure 386   ♀ dorsal

Figure 386   *Neophyllobius agrifoliae* McGregor

Patellae bearing two visible setae, shorter than segment; dorsum striae mostly transverse; 17 pairs of dorsal setae, including caudal marginal setae, arranged as four submarginally along front of body, seven sublaterally each side between coxae I and caudal margin, six submedium pairs between coxae II and hind margin, 4 along caudal margin; legs longer than body; tarsi swollen, shorter than tibiae, each bearing 2 strong simple claws, with pulvillus in center associated with tenant hairs.

Collected in California on live oak (*Quercus agrifolia*).

♀ dorsal

Figure 387

Figure 387 *Neophyllobius lombardinii* Summers and Schlinger

Dorsal setae beset with coarse, diverging spikelets, length of spikelets exceeding diameter of setae; three pairs of long dorso-median setae not tapered to points, basal tubercles of each pair situated side by side, almost confluent on their median surface; femur I bears four setae about equal distance one from the other; femur IV with two setae of equal lengths; 14 pairs of dorsal setae, arising on prominent tubercles; sensilla on genu I and II and tibia I to IV, a claviform sensillum proximally on tarsi I and II.

Collected in California within oak leaf mold.

Family Syringophilidae

This is a family that is found associated with birds and has been called quill mites. They are characteristically found living inside the shafts of the large flight feathers of the wings of birds. However, they may occur in the tail feathers or even the smaller body feathers. For many years the only species commonly referred to within this family was *Syringophilus bipectinatus* and *S. columbae* associated with poultry and domestic pigeons. In the United States the work of Clark (1964) is the only comprehensive treatment of these mites. Syringophilids are very elongate forms, with long, slender, piercing stylets that allow them to feed on the soft vascular living tissue within the feather quill.

The Syringophilids are not the only mites that may be encountered living in the quills as members of the large superfamily Analgoidea also utilize the quill as a protective habitat. Seven species of this faimly have been described and recorded by Clark (1964) for the United States. This family contains only the single genus *Syringophilus*.

Small elongate, thin, worm-like, milky white or opaque yellow mites; body divided by indistinct constriction into three main regions; peritremes well developed; setae normally long and hair-like; legs well developed .... **Genus *Syringophilus***

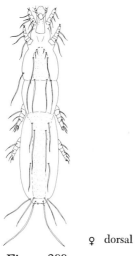

Figure 388   ♀ dorsal

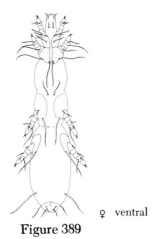

Figure 389   ♀ ventral

Opisthosoma with prominent, punctate dorsal plate extending anteriorly to mesosomal constriction; dorsal setae weakly spinose at margins; rostrum unornamented; propodosomal dorsal shield sclerotized; opisthosomal shield moderately sclerotized; palpal tarsus with claw-like and thumb-shaped setae; rostrum with bilateral peritremes dorsally, and bearing a sclerotized, punctate plate dorsally.

Collected from Maryland on *Seiurus aurocapillus* (oven bird), *Helmintheros vermivorus* (eating warbler), *Melospiza melodia* (song sparrow).

Family Scutacaridae

The family Scutacaridae has been characterized as bizzare creatures by Baker and Wharton (1952). Hughes (1961) refers to the small number of genera described within the family and its wide distribution associated with moss and humus. The species *Acarapis woodi* is the best known member of the family and is found in the tracheae of bees. This acarine disease of

bees (also known as the Isle of Wight disease) was so important that the United States Congress placed restrictions on the importation of queens and worker bees from foreign countries. This mite is not present in the United States but has been recorded from Argentina as well as Europe, Russia, the British Isles. Krantz (1970) does not utilize the family name Scutacaridae in his treatment of the superfamily Tarsonemoidea, but he does use it in his key to the families of the Prostigmata. Mahunka (1964) is followed in this work and whether the Scutacarids should be considered as a subgroup within the family Pyemotidae remains for future works to decide.

1a   **Last two joints of leg IV, tibia and tarsus, distinct, leg 5-jointed; however, when praetarsus present, leg 6-jointed; tarsus of leg IV similar in shape to other legs; legs short, not attaining posterior margin of body; inner margin of trochanter of leg IV usually with a pointed, spur-like protuberance** .... **Genus** *Pygmodispus*

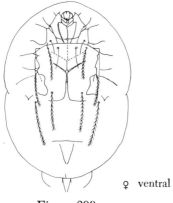

Figure 390   ♀ ventral

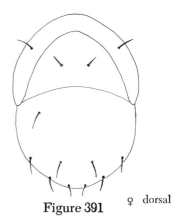

Figure 391   ♀ dorsal

Figures 390 and 391   *Pygmodispus calcaratus* Paoli

Trochanter IV with a large, sharp, strongly projecting spur on inner margin; poststernal setae of equal length, arising far from one another, praesternal setae internal, originating in front of external seta; epimere III obliquely proclinate.

1b   **Last two joints of leg IV, tibia and tarsus, fused into a tibiotarsus, four-jointed, no ambulacrum terminally, six or seven hairs on tibiotarsus; external humerals arising only on clypeal margin** .... **Genus** *Scutacarus*

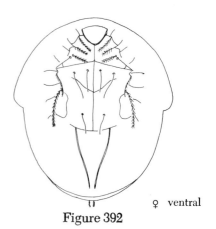

Figure 392   ♀ ventral

Figure 392   *Scutacarus crassisetus simplex* (Paoli)

Sacral setae external, glabrous, lumbal setae, external, short, thin; dorsal hairs finely and sparsely ciliate; tibiotarsus of leg I small ending in a claw; some dorsal hairs incrassate or brush-like; internal lumbales and sacrales equal in length, external lumbales and sacrales unequal.

Keys to the Common Astigmatids of the Astigmata

1a Free-living in the soil, on plants, in association with animals and their nests; usually soft-bodied; with two pairs of distinct genital discs or ringlike structures; caruncle and empodial claw present on some or all tarsi ....
supercohort Acaridia **2**

1b Parasitic either as ectoparasites or endoparasites on birds, mammals, crustaceans or insects; may be soft-bodied or with sclerotized plates; genital discs greatly reduced or absent; caruncles suckerlike or absent; apex of tarsi with clawlike setae or spines ....
supercohort Psoroptidia **13**

2a Female genital aperture a transverse slit; adults and deutonymph free-living in moist habitats; adult female discs not asociated with genitalia; caruncles reduced; deutonymph sucker plate well developed; legs III-IV more slender than I-II (p. 252) .... Family Anoetidae

2b Female genital various, but never a transverse slit; deutonymph with legs III-IV more or less equally formed **3**

3a Tarsi with narrow suckerlike caruncles distinctly connected to tarsus by stalk; genital and anal opening of female contiguous and nearly continuous in male, at posterior end of idiosoma. Predators of scale insects ....
.... Family Hemisarcoptidae

This family contains only a few species, three according to Krantz (1970). Two of these, *Hemisarcoptus malus* (Shimer) and *H. cooremani* (Thomas) have been reported from the United States. *H. cooremani* is known only from the hypopus collected in Texas on *Chilocorus cacti*. *Hemisarcoptus malus* has been collected from Massachusetts, Ohio, Illinois, Iowa and South Dakota and is always found in association with the oyster-shell insect, *Lepidosaphes ulmi*.

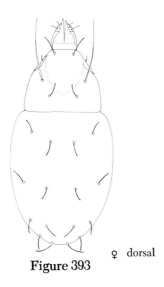

Figure 393

♀ dorsal

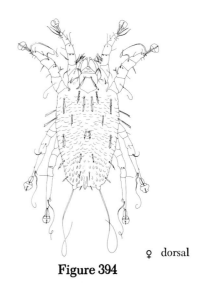

Figure 394

♀ dorsal

Figure 393 *Hemisarcoptus* sp.

3b **Tarsi with caruncles broadly connected to tarsus, not stalked, rarely suckerlike, empodial claws well developed, distinct** **4**

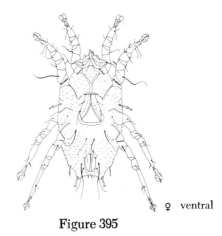

Figure 395

♀ ventral

4a **Pseudorutellar process bilobed, membranous, located on mediodistal aspect of subcapitulum; two setae (one median and one lateral) at base of fixed cheliceral digit. Associated with bats** .... **Family Rosensteiniidae**

This family was placed as a synonym of the family Canestriniidae by Baker et al., (1958), McDaniel and Baker (1962) reevaluated the family and considered the Rosensteiniidae Cooreman (1954) as a distinct family and described a new genus and species for the family from Mexico. Strandtmann (1962) described a new species of mite in the family Glycyphagidae. This species belongs to the family Rosensteiniidae and is recorded from the United States. .... Genus *Nycteriglyphus*

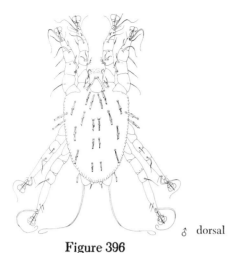

_{♂ dorsal}

Figure 396

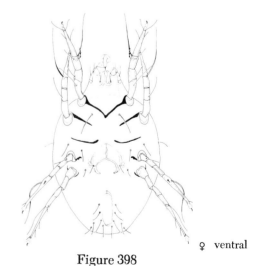

_{♀ ventral}

Figure 398

genitalia between coxae IV; female genitalia between coxae II and III; setae sa i smooth as long as body.

Collected in cave in association with *Tadarida brasiliensis* from Texas.

4b   Pseudorutellar process absent, single seta on median aspect of chelicara 5

5a   Two sclerotized lateral plates cover genital opening of female forming a crescent; dorsal body setae minute; male with tarsal and anal suckers; genital opening betwen coxae I-II closest to coxae I .... Family Chortoglyphidae

Mites which make up this family in the United States are associated with barns and granaries. They are known to be associated with flour, wheat, rye, oats, grass seed, clover seed, and heaps of old straw in many parts of the world. Krantz (1961) recorded *C. arcuatus* as a granary mite known to occur in stored grain and grain products in the Pacific Northwest. .... Genus *Chortoglyphus*

Figure 397

Figures 394, 395, 396, and 397 *Nycteriglyphus bifolium* Strandtmann

Apodemes I united; body covered with scale-like wrinkles; dorsal setae flattened, serrated or fringed; propodosomal shield present; male without copulatory anal or tarsal suckers; female genitalia between coxae II and III; dorsal setae with three or four serrations; supracoxal seta two-tined; genu I with paired solenidia; ventral setae smooth, whiplike, male

Figure 398 *Chortoglyphus arcuatus* (Tropeau)

Idiosoma dorsal setae very short, smooth, scarcely visible; cuticle smooth; setae of tibiae and genu I and II pectinate; setae arising from ventral surface of tarsus spinelike; male with tarsal and anal suckers; genital opening strongly displaced anteriorly, situated between ends of epimeres I; chelicerae large, distinctly toothed; propodosomal shield absent; body setae short and fine, "Vi" projects forward over gnathosoma, "Ve" longer than "Vi"; four pairs of dorsal setae; male genital opening located between coxae I and II; apodemes I and II forced apart and form part of transparent genital folds; no sternum present; legs long and slender; female apodemes I unite to form a short sternum, apodemes II are long, an elongated sclerite lying between coxae II and III.

Collected in stored grain from Oregon, Washington, Idaho and New York.

Remarks: Banks (1915) recorded *Chortoglyphus gracilipes* Banks from tobacco. Zakhrakin (1941) referred to *C. gracilipes* Banks as the common North American form of the genus *Chortoglyphus*. The description of *C. gracilopes* is too general to establish its separation from *C. arcuatus*.

5b   Lateral plates of female not strongly delineated, not forming a crescent; genital discs well developed; male genital opening normally between coxae III-IV
.... **6**

6a   Adults with two sclerotized rods (condylophores) connecting empodial claw to pretarsus; males usually with tarsal and anal suckers; both sexes weakly sclerotized; deutonymphs with sucker plate and gnathosoma well developed; terminal sensory organ (seta omega)

elongate, exceeding length of subcapitulum; sternal setae often modified into discs (p. 238) .... Family Acaridae

6b   Adults with at the most a single tendon (not condylophores) connecting empodial claw to pretarsus; males normally without tarsal or anal suckers **7**

7a   Tarsi I-II short, ending in a short process at base from which arises a fine, long, flexible pretarsus with small claw, pretarsi III-IV short; idiosoma distinct; associated with barnacles or algae in a marine habitat ....
.... **Family Hyadesiidae**

Members of this family are found associated with tide pools and submerged algae, Manson (1963) in describing a species found in California, placed it in the family Carpoglyphidae. .... Genus *Hyadesia*

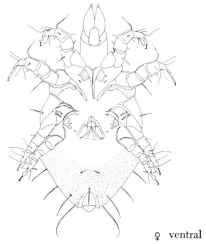

♀ ventral

Figure 399

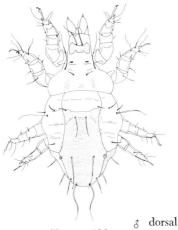

♂ dorsal

Figure 400

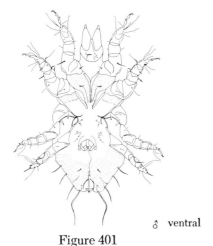

♂ ventral

Figure 401

Figures 399, 400, and 401 *Hyadesia glynni* Manson

Tarsi I and II short, ending in a stout process at base where arises a fine, long, flexible pretarsus, with a small claw, pretarsi III and IV short; idiosoma with a distinct transverse constriction; dorsal shield covers greater part of hysterosoma extending over posterior portion of ventral surface; suture between propodosoma and hysterosoma distinct; apodemes I join sternum, apodemes III and IV joined; tarsi I and II terminating in large claw, pre-

tarsus flexible, long, broadened, distally; female tarsus I with eight setae, tarsus II with seven setae, tarsi III and IV with five setae, tibia I and II with stout conical spine.

Collected in association with barnacle, *Balanus glandula* from California.

7b **Without long flexible pretarsus not attached to distal extremity of tarsus, non-marine in habitat** 8

8a **Idiosoma fusiform, narrowing sharply, cuticle smooth; external scapular setae capilliform; dorsal setae consisting of microsetae; legs I strongly compressed laterally; dark brown, associated with nests of rodents** .... **Family Fusacaridae**

Members of this family are associated with the nests of small mammals. Only the single species, *Fusacarus laminipes*, is recorded from the United States. .... Genus *Fusacarus*

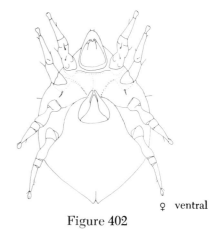

♀ ventral

Figure 402

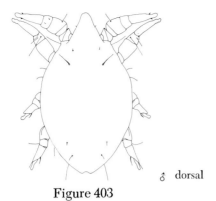

Figure 403

δ dorsal

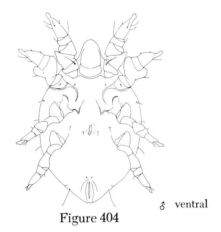

Figure 404

δ ventral

Figures 402, 403, and 404 *Fusacarus laminipes* Michael

Idiosoma fusiform, narrowing sharply, cuticle smooth; external scapular setae capilliform; dorsal setae consisting of microsetae; legs short, legs I strongly compressed laterally; male epimeres II curved and free, epimeres III and IV joined, the junction being produced internally, bent at a right angle; male genital opening placed between bases of legs III and IV; female genital opening between legs II and III, surrounded by sclerotized ring, both male and female with idiosoma fusiform, convex, integument smooth; thick, dorsal setae very small except for external scapular setae.

Collected from the nests of rodents in the eastern section of the United States.

8b **Idiosoma various but not as above, dorsum with more than single pair of macrosetae** **9**

9a **Idiosoma edged with 11-12 pairs of bipectinate setae, leaflike in shape; vertical, scapular and dorsal setae similar to peripheral setae; males quite different in shape from female; dorsal setae long, relatively narrow, palmate or leaflike ....** .... **Family Ctenoglyphidae**

Ctenoglyphids are found associated with stored products, and are commonly found in the dust of hay and fodder in warehouses. Zakhvatkin (1941) recorded four species in the genus *Ctenoglyphus*. The species *Ctenoglyphus plumiger* Koch has been recorded from Oregon .... Genus *Ctenoglyphus*

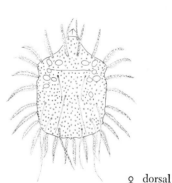

$\female$ dorsal

Figure 405

δ dorsal

Figure 406

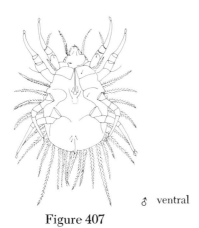

♂ ventral

Figure 407

Figures 405, 406 and 407 *Ctenoglyphus plumiger* (Koch)

Idiosoma edged with 11 to 12 pairs of pectinate setae, leaflike in shape; vertical, scapular and dorsal setae similar to peripheral setae; males quite different in shape than female; male dorsal setae long, relatively narrow; female dorsal setae palmate or leaflike; male with dorsal setae finely pectinate, each with numerous fine barbs, dorsal setae one shorter than dorsal setae two; female with all dorsal setae narrow, finely pectinate, barbs of the bipectinate setae straight, forming an acute angle with axis of seta; cuticle of dorsal surface covered with cylindrical warts, with ends which are split, forming a rosette; male with apical seta of tibia I very long.

Recorded by Krantz (1970) from Oregon.

9b   Idiosoma not edged with 11-12 pairs of bipectinate setae that are leaflife in shape 10

10a   Idiosoma divided by a sejugal furrow, between propodosoma and hysterosoma in adults, body setae smooth, female genitalia relatively large; small mites;

free-living (p. 223) .... Family Saproglyphidae

10b   Idiosoma without a complete transverse sejugal furrow between propodosoma and hysterosoma in adults; (deutonymphs may have a sejugal furrow; however, empodial claws are sessile and pretarsi not long and stalklike as in deutonymphs of Saproglyphidae) 11

11a   Claws distinctly larger than caruncles, often half the length of tarsus, associated with Hymenoptera .... Family Chaetodactylidae

Wood-boring bees or their nests are where members of the family Chaetodactylidae are found. The bees provide a method by which these mites are able to migrate from old nests of these bees to new nests. The hypopi is the usual migrating stage while the adults are known to feed on the eggs of their transporting agents, the bees. A worker interested in these mites should consult the work of Krombein (1962 ab).

11b   Claws small, if well developed, not larger than associated caruncle; not commonly associated with Hymenoptera 12

12a   Coxal apodemes I-II of female thickened, fused to sternum and bordering anterior edge of genital shield; tarsi never more than twice the length of adjacent tibia; associated with dried fruit in the United States .... Family Carpoglyphidae

Members of this family were considered a subfamily of the Glyphagidae by Zakhvatkin (1941). They are similar to the Acaridae .... Genus *Carpoglyphus*

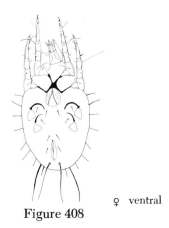

♀ ventral

Figure 408

Figure 408 *Carpoglyphus lactis* (L.)

Cuticle smooth; tarsal and anal suckers absent in male; apodemes of legs II join the sternum; propodosomal shield present, no transverse groove; gnathosoma narrow; idiosoma flattened; most dorsal setae about equal in length with rounded apex, short, smooth; external scapular setae slightly longer than other setae; genital folds of female well developed; epimeres III and IV free, wide apart at apex; legs I and II of male thicker than same legs of female; genital opening of male placed between bases of legs III and IV without sclerotized ring; apodemes well sclerotized with those of leg I joining to form a sternum which forks at its distal end and articulates with apodemes of legs II; two pairs of genital setae are present in the male; anus extends as far as the posterior end of body.

This is a cosmopolitan species associated with dried fruit.

12b Coxal apodemes I-II of female various, not thickened and fused to sternum; tarsi various, often much longer than length of adjacent tibia; integument strongly pigmented, occasionally spined; legs I-II often compressed laterally or ornamented with longitudinal keels; genital sclerites well developed, often

forming a circumgenital ring (subfamily Labidophorinae); integument soft, colorless, smooth, wrinkled; legs rarely ornamented or compressed (subfamily Glycyphaginae (p. 231) .... Family Glycyphagidae

13a Genital aperture of female a simple transverse slit without genital apodemes; body rounded or saclike; some males may be elongate; legs of females often reduced, skin parasites of vertebrates 14

13b Genital aperture various, never a simple transvere slit; genital apodemes present, may be reduced (Turbinoptidae); idiosoma elongate or ovate; legs not reduced, with terminal or subterminal empodial suckers 15

14a Prodorsum with strong elongate lateral apodemes which flank or encircle the prodorsal sclerite; parasites of fowl and small domestic birds (p. 283) .... Family Knemidocoptidae

14b Prodorsum without elongate lateral apodemes; parasites of mammals (p. 270 .... Family Sarcoptidae

15a Pedipalpi modified into two lateral tubes which enclose the simple chelicerae; legs without suckers, legs III-IV modified and enlarged for grasping gills or hairs; external parasites of shore crabs .... Family Ewingidae

This family contains only the single species, *Ewingia cenobitae* Pearse, and is found associated with the gills of pagurid crabs. .... Genus *Ewingia*

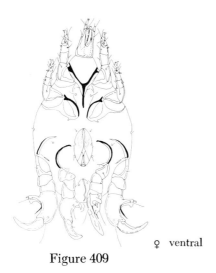

♀ ventral

Figure 409

Figure 409 *Ewingia cenobitae* Pearse

Legs III and IV modified into claspers; pedipalpi modified into two lateral tubes which enclose the simple chelicerae; female with three pairs of setae on dorsal surface, middle pair lateral, other two pairs nearer median line; a pair of small setae placed posterior to anus; genital pore between coxal plates of legs III; capitulum, conical less than half the length of remainder of body, basal segment slightly wider than long; pedipalpi two-segmented, with three teeth on each proximal half of the median surface of distal segment; chelicerae three-segmented, flattened, linear, median edge, thin, sharp, two teeth at tip at distal segment at median angle; legs I and II stout; suckers absent; legs III and IV chelate, penultimate segment bears toothed process on anterior margin.

Collected near the outside of the gills of *Cenobita diogenes* off loggerhead key in Florida.

15b **Pedipalpi not modified into two lateral tubes that enclose chelicerae; legs with or without suckers, legs III-IV may be modified and enlarged for grasping**

hairs but never in combination with a tubular pedipalpi **16**

16a Genital opening of female a longitudinal slit, if opening appears transverse, it is located between coxae II or the slit may widen at its posterior apex to form an inverted "V"; subdermal, visceral and lung parasites of mammals and birds **17**

16b Genital opening various but never a simple longitudinal slit, transverse opening, between coxae III or IV **19**

17a Body elongate; legs I-II thick, conical, short, femur and genu coalesed dorsally, legs IV inserted on anterior portion of idiosoma; internal parasites of crows and domestic fowl .... **Family Laminosioptidae**

Species of this family are known as subcutaneous parasites of birds and are called the fowl cyst mites. Work done in Michigan by Lindquist and Belding (1949) recorded that small cysts were discovered in the skin of the neck, breast, flank, vent, and thighs of birds. There is at present only two species recorded from the United States, both within the genus *Laminosioptes* .... Genus *Laminosioptes*

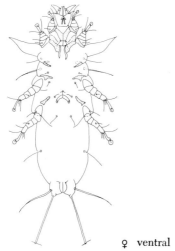

♀ ventral

Figure 410

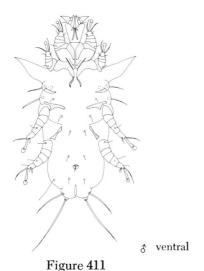

♂ ventral

Figure 411

Figures 410 and 411 *Laminosioptes hymenopterus* Jones and Gaud

Legs IV inserted on anterior portion of idiosoma, femur and genu I and II each coalesced dorsally; body cylindrical, tapered on both ends; propodosomal shield finely granulated, lateral borders outlined by chitinous thickening pronounced in posterior region; pair of external and a pair of internal scapulary setae,

located on outside of posterior border of propodosomal shield; tarsi I and II with caruncles; a pair of membranous alae near maxillary palps and on propodosoma; abdomen bears two pairs of long terminal setae.

Collected on *Corvus brachyrhynchos* from Ohio.

17b Body oval, egg-shaped; legs I-II various, but not thickened, femur and genu not coalesed dorsally, legs IV inserted on posterior portion of idiosoma **18**

18a Coxae of legs forming large anterior and posterior fused ventral sclerites; tarsi each with a long seta; parasitic in lungs of rodents (p. 262) .... **Family Pneumocoptidae**

18b Coxae of legs not greatly enlarged or fused, tarsi without long seta; internal parasites of fowl .... **Family Cytoditidae**

Only a few species of this family have been collected throughout the world. Fain and Bafort (1964) recorded nine species within the family Cytoditidae in their work. Hyland (1969) described a new species, *Cytodites therae*, and stated that except for *C. nudus* (Vizioli), *C. therae* is the only record of this family in North America.

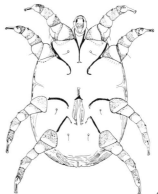

♀ ventral

Figure 412

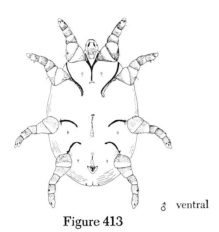

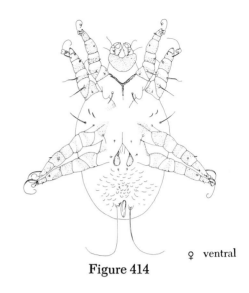

Figure 413

♂ ventral

♀ ventral

Figure 414

Figures 412 and 413 *Cytodites nudus* (Vizioli)

Female tarsi with all pedunculate ambulacra of legs I-IV the same size; epimera III and IV not fused to form a continuous chitinous arc.

Collected in lungs of *Bonasa umbellus, Meleagris gallopavo, Gallus domesticus* throughout the United States where hosts are found.

19a **Parasites of respiratory passages of rodents or birds** **20**

19b **If parasitic, not associated with respiratory passages** **21**

20a **Integumentary striae of soft cuticle, composed of scalelike plates; apotele suckerlike; adanal copulatory suckers absent in male; parasites of respiratory passages of rodents** ....
.... **Family Yunkeracaridae**

The family Yunkeracaridae contains only a single genus, *Yunkeracarus* Fain and two species, one of which has been recorded from the United States. The genus was at one time included within the family Epidermoptidae.

Figure 414 *Yunkeracarus faini* Hyland and Clark

Propodosomal setae present, median pair longer than laterals; epimera I fused into "Y"; genital apodeme of female small, located at level of coxae III, opening an inverted "Y," associated with two microsetae on each side; all body setae small; surface of opisthosoma clothed with small transparent scales; chelicerae well developed, strongly chelate; tarsi I and II with short whip-like setae.

Collected in nasal cavity of the white-footed mouse, *Peromyscus leucopus* from Michigan and Texas.

20b **Integumentary striae without scalelike plates; tarsi I-II markedly shorter than tibiae I-II; male with copulatory suckers, often bilobate posteriorly; parasites of respiratory passages of birds** ....
.... **Family Turbinoptidae**

This family contains only two species recorded from the United States associated with birds. The two species are presently placed in separate genera, these being *Turbinoptes* and *Schoutedenocoptes*.

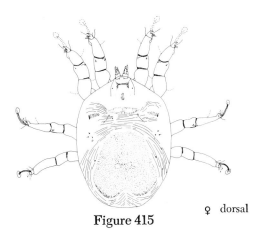

Figure 415   ♀ dorsal

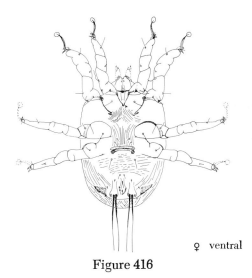

♀ ventral

Figure 416

Figures 415 and 416   *Turbinoptes strandt-manni* Boyd

Female vulva transverse; eyes lacking; copulatory suckers absent, pair of prominent copulatory cones present, lateral to anus; genital aperture between coxae IV; epimera I united in "V" shape; body setae hairlike, dorsum with six pairs, three small lateral setae grouped together; venter with 11 pairs of setae, genital setae arranged as two anterior, one posterior; anal setae arising on copulatory cones; male with copulatory suckers; genital aperture posterior to coxa IV with prominent short curved penis.

Collected in nasal cavity of ring-billed gull *(Larus delawarensis)* from Texas.

21a   **Propodosoma with a rigid bifurcate or trifurcate epistomal extension; legs III-IV of female inserted ventrally; males with or without anal suckers; nidicoles in nests of mammals or birds, (subfamily Pyroglyphinae); free-living house dust mites which cause allergic asthma and dermatoses of humans, (subfamily Dermatophagoidinae) (p. 257)** .... **Family Pyroglyphidae**

21b   **Propodosoma without a rigid epistomal extension; not free-living forms** **22**

22a   **Parasites of the skin of mammals, birds, (may be associated with parasitic insects of birds, notably the Hippoboscidae and Mallophaga** **23**

22b   **Parasites of the fur of mammals or the feathers of birds** **24**

23a   **Skin parasites of mammals; epistomal portion of propodosoma truncate or rounded; legs III-IV of female inserted laterally or ventrally (when inserted laterally, often reduced and with long whiplike terminal setae); males with anal suckers (p. 228)** .... **Family Psoroptidae**

23b   **Skin parasites of birds, or associated with insects that are parasites of birds; body shields weakly developed; apoteles of tarsi bell-shaped or rounded, tarsi of-**

ten clawlike distally; vertical setae absent; palpi often with flanges or enlargements (p. 271) .... **Family Epidermoptidae**

24a External parasites of the fur of mammals, commonly called fur mites .... (superfamily Listrophoroidea) 25

24b External parasites of the feathers of birds, commonly called feather mites .... (superfamily Analgoidea) 28

This group of families is known to acarologists as "feather mites" and belongs to the superfamily Analgoidea. Atyeo and Braasch (1966) have published one of the most extensive studies that treat members of the family Procotophyllodidae in their *"The Feather Mite Genus Proctophyllodes."* Anyone attempting to identify members of this family should consult the works of Dubinin (1951, 1953, 1956), Gaud and Mouchet (1957-1959) and Gaud (1959-1974). It must be pointed out that in identification of members of this superfamily a worker should not attempt to establish the identity on the basis of its host, as the host bird may well spend its winter in South America or even Africa. This points out the need in consulting the literature of workers studying feather mites in other parts of the world. An example to illustrate this is the genus *Proctophyllodes* given by Atyeo and Braasch (1966) in which they cited that 38 percent of the members of this genus are known from a single host, yet *P. reguli* has been reported from French Morocco, England and the United States, from three different hosts.

If one enjoys working with unknown specimens, and is willing to work from publications written in Russian, German, French, and Spanish, the members of the superfamily Analgoidea will more than challenge the beginner and specialist alike. As Atyeo and Braasch (1966) so aptly pointed out, with approximately 8600 species of birds in the world, a worker can be assured of rich resources for future investigations.

25a Tegmen well developed, scleritized, concave ventrally and completely covering the area in front of gnathosoma; sternal region with two large striated membranes stretching over ventral surface of gnathosoma; leg with five free segments, tarsus terminated by a sucker; male with adanal suckers (p. 264) .... **Family Listrophoridae**

25b Tegmen absent not covering area in front of gnathosoma 26

26a Legs I and II normally developed not modified for clasping fur, legs III and IV of the female and legs III of the male pincerlike for clasping hair of its host; body dorsoventrally flattened (p. 280) .... **Family Myocoptidae**

26b Legs I and II modified for clasping fur, legs III or IV not pincerlike 27

27a Anterior legs thickened and modified at their extreme anterior apex, but with normal number of leg segments, ambulacase small, deprived of membranes, posterior not pincerlike, tibia united to corresponding tarsi; body dorsoventrally flattened or subcylindrical (p. 269) .... **Family Atopomelidae**

27b Anterior legs deprived of ventral ambulacases and with either five leg segments (subfamily Chirodiscinae) or reduced to one or two segments (subfamily Labidocarpinae); posterior legs either normal or with genu united to femur, tarsi and tibia always movable segments; with chitinized sternal membranes; legs I and II in the subfamily Labidocarpinae short modified but open apically to form a large chitinized striated membrane, recurved inside to clasp hair of host; body of this subfamily normally compressed laterally; body of subfamily Chirodiscinae dorsoventrally flattened. (The latter subfamily contains a single species *Chirodiscus amplexans* from Africa) (p. 277) .... Family Chirodiscidae

28a Legs arising from margins of body; coxae III-IV visible from dorsal portion of body; coxal apodemes only rarely fused 29

28b Legs III-IV inserted ventrally, not projecting beyond margin of body; coxal apodemes often fused together, resulting in closed coxal fields; apoteles large normally oval or heart-shaped (fig. 417) .... Family Freyanidae

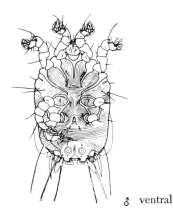

♂ ventral

Figure 417

Figure 417 *Freyana anatina*

Representatives of this family are associated with water fowl and marsh birds. The males are ordinarily polymorphic in body structure. Both sexes have a broadly oval form with the post anal and sacral setae usually modified into leaflike or otherwise-shaped lamellae or plates. Many of the species have some setae modified for the purpose of attachment onto a feather. These setae may be either body or leg setae. In the genus *Michaelichus* some males have the setae located on the base of the tarsus of the second left leg hooked for catching onto the barbule of the host's feather. In the genus *Freyanella,* large modified scapular setae on the right side of the body are set against the wall of the opposing barbule and tightly wedge the mite against it .

Within the mites of this family the coxosternal skeleton sustains its greatest development. In the males, all elements of this skeleton are fused together and form an almost continuous network of sclerites on the ventral surface. In females this coxo-sternal skeleton is weakly developed. The body is asymmetrical in shape in most of the species within the family Freyanidae. With the asymmetrical development of the legs, the individual epimeres and their coxal fields are more developed, bringing about an abruptly broken symmetry of the body.

29a Transparent "spurs" and short stout setae on tarsi, tibiae or genu of legs I-II; coxal apodemes III-IV of male forming coxal plates; legs III of male normally much longer and thicker than same legs of female (figs. 418, 419, 420) .... Family Analgidae

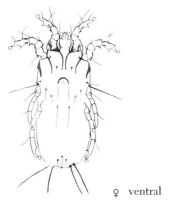

♀ ventral

Figure 418

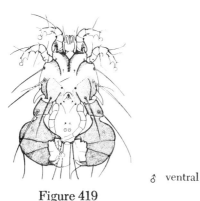

♂ ventral

Figure 419

Figures 418 and 419 *Analges*

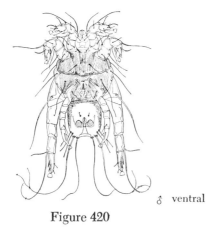

♂ ventral

Figure 420

Figure 420 *Megninia*

Members of this family are parasitic on almost all orders of birds. The chief place of localization appears to be on the small covert feathers of the wings and body. Species of the genus *Megnina* and *Mesalges* live on the proximal primary and secondary wing feathers and on the rectrix feathers, more often in their bases, but never inhabiting the primary wing feathers distally.

In this family, most genera are diagnosed by the males having legs either greatly enlarged or reduced. In the two largest genera, such as *Megnina* and *Mesalges*, the males are characterized by having the third pair of legs enlarged. However, contrasting this is the genus *Xolages* in which the fourth pair of legs are more developed. In other members both posterior pair of legs may be greatly enlarged. The first pair of legs contains a chitinous outgrowth in the form of a hook which has been utilized by some workers to characterize the members of this family. The males, in the majority of the genera, have large opisthosomal lobes; the edges of which are enclosed by a hyaline emargination. Legs I, II and III are also found to articulate at the edge of the body; legs IV may frequently be submarginal.

29b   Transparent "spurs" and short stout setae on tarsi, tibiae or genua absent; legs III of male similar in size and shape as female 30

30a   Posterior end of body of female (and the majority of males) bilobed, these lobes may be secondarily fused and form a narrow "tail-like" appendage; opisthosomal lobes with long setae often modified and having hyaline blades which may be sword or leaflike in shape (figs. 421, 422, 423, 424) ....
.... **Family Proctophyllodidae**

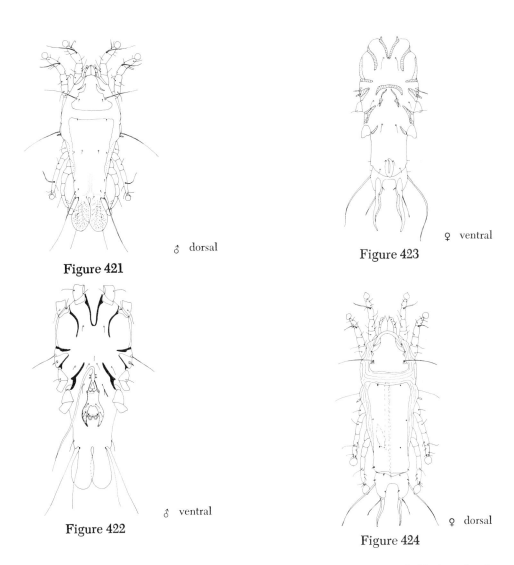

Figure 421   ♂ dorsal

Figure 422   ♂ ventral

Figure 423   ♀ ventral

Figure 424   ♀ dorsal

Figures 421-424   *Proctophyllodes glandarinus*

The family Proctophyllodidae, according to Atyeo and Braasch (1966), contains twenty-four genera within three subfamilies. The body of members of this family are elongated, narrow, with almost parallel sides. The female is distinctive in having two opisthosomal lobes on the posterior end of the body. These lobes contain setae that are often expanded in a sword- or leaflike manner. The males also have two lobes on the posterior apex of the body. However, there are exceptions in the male as these lobes may coalesce into a caudal appendage as found in members of the genus *Alloptes*. The dorsal shields are well-developed, containing small punctures. Ventrally the coxo-sternal skeleton appears quite variable; anal copulatory suckers are well-developed with the genital aperture between the bases of legs IV, rarely legs III. The structure of the male copulatory organ is highly variable; the character of this structure and its associated sclerites present good diagnostic characters for distinguishing different genera and species, thus pointing out the importance of collecting males. Many males possess a very long aedeagus which is distinctive.

The hosts of this family range throughout many families of birds. Some genera such as the *Proctophyllodes* are usually parasites of birds of the order Passeriformes. Members of this family are to be found along the rachis or on the remiges of the wings and tail feathers. The mites position themselves in tandem adjacent to the barbs of the rachis.

According to Atyeo and Braasch (1966), members of the genus *Proctophyllodes* are scavengers and do not cause any serious damage to their bird hosts. This is probably true for all members of this family.

30b   **Posterior end of body of female and male broadly rounded, rarely bilobed, never with opisthosomal lobes, never with sword or leaflike appendages (figs. 425, 426) .... Family Dermoglyphidae**

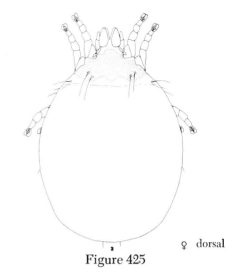

♀ dorsal

Figure 425

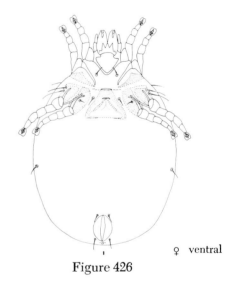

♀ ventral

Figure 426

Figures 425 and 426   *Sphaerogastra crena*

This is a large family and now includes the pterolichids once placed in a family with the Dermoglyphidae as a subfamily. There is no single structural form that can be utilized to distinguish all the members of this family. They are usually large, with a thick body, mature female nymphs and larvae have a rounded posterior abdominal region, sometimes with a slight emargination, but never lobed as in the family Proctophyllodidae. The males however, in many genera may have deep emarginations, deliniated by two opisthosomal lobes. In both sexes the propodosomal shield is well-developed and may occupy the entire dorsal surface of the propodosomal.

The coxo-sternal skeleton is well-developed in that the epimeres of legs I in some species fuse to form a "V"- or "Y"-shaped sternum. However, in contrast to the family Freyanidae, the other epimeres as a rule are separate from one another. Also, the legs articulate laterally. The segments of the legs are cylindrical and without membranelike expansions found on members of the family Analgidae.

Representatives of this family live for the most part on the wing feathers, rarely on the body. Many are found within the quills of the wing feathers. However, they are not the only group of mites that may be found inhabiting the inside of a quill as members of the family Syringobiidae are also found occupying this microhabitat.

Family Saproglyphidae

Saproglyphid mites are cosmopolitan. Many infest the bodies of insects and do not assume the role of a parasite. Many are known only through the hypopial life stage. As adults and immatures they have functional mouthparts and may be free-living or associated with dead, decaying, organic materials. The members of this family occupy all kinds of habitats both indoors and out. They are reported to survive best in high humidity and low oxygen content and tend to avoid the direct rays of the sun.

There are species, such as *Calvolia romanovae*, that have been recorded from stored agricultural products. *C. zacheri* is a pest of moldy cheese and *Saproglyphus neglectus* is associated with mushrooms and toadstools.

By far the largest number of described species are associated with insects such as wasps, bees, flies and beetles. Most of the work in the United States has been with the hypopi that are found on wasps. This stage of the mite lacks mouthparts and has a ventral plate equipped with suckers which aid in attaching to an insect host. A considerable amount of work has been done regarding the biological relationship between wasps and the phoritic mite hypopi. The symbiotic mite *Di-*nogamasus braunsi, lives in the abdominal pouches of the female yellow-banded carpenter bees until nesting begins. It then leaves the pouches to enter the bee nest where the female mite lays eggs upon the pupae of the developing bee.

Krombein (1961) and Mostafa (1967) found the hypopi, when leaving the pupal eruviae, will climb onto an adult wasp, crawl around until it reaches the apical margin of a specialized mite chamber called an acarinarium found at the base of the second abdominal segment of the wasp. The hypopi then turn around and back into the acarinarium. It has been observed that as additional hypopi perform this same maneuver, they come to lie in single rows with their anterior ends facing the outside opening of the chamber. In this manner the hypopi are then ready to infest a new nest and resume their life cycle.

Certain sparoglyphid mites have adjusted their developmental cycles perfectly to those of their hosts. The work of Krombein (1961) has shown that the rhythmic pulsation of the female wasp's abdomen during oviposition may stimulate the hypopi of the genus *Vespacarus* to forsake the body of the wasp and drop into a new uninfested nest or cell.

It was established that the transformation of the hypopi first to tritonymphs and then to adults takes place in the interval between oviposition of the female wasp and the completion of feeding by the wasp larva.

It is believed that species of the genus *Vidia* associated with the Douglas fir beetle feed on nematodes occurring under the elytra of the beetle thus the association may directly benefit both the mite and the beetle.

In attempting to work with this group an individual should consult the works of Mostafa (1970), Krombein (1961-2), Baker (1962-4) and Baker and Cunliffe (1960).

1   Coxal apodemes IV meet mesally forming a central junction that extends posteriorly into metasternum IV; internal gnathosomal setae long; tarsi III and IV without digitiform process; claws of tarsi I to III small; apex of tarsus pointed, narrow toward its tip; eyes present .... **Genus *Vespacarus***

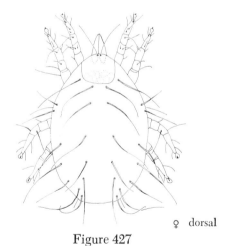

Figure 427   ♀ dorsal

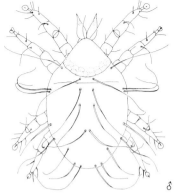

♂ dorsal

Figure 428

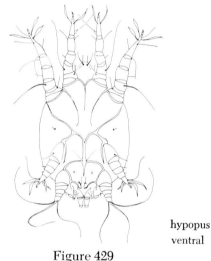

hypopus ventral

Figure 429

Figures 427, 428, and 429   *Vespacarus histrio* Baker and Cunliffe

Hypopus with apodemes III not contiguous; sternum IV flanged; anterior part of propodosoma triangular; eyes less than eye width apart; apodemes III strong, well delineated medially; junction of apodemes IV projecting only slightly anteriorly, the two arms connecting apodemes III and IV barely visible; rod-like sensory seta on tibia IV small; propodosomal shield about as wide as distance between outer propodosomal setae; male dorsal setae with the exception of anterior propodosomal, strong, whiplike.

Collected in association with *Stenodynerus histrio* and *S. fulvipes fulvipes* from North Carolina.

2 Tarsus IV blunt distally; proximal leaf-shaped seta on tarsi I and II with slender stalk; pore on coxal field I faintly outlined; terminal setae developed .... **Genus *Vidia***

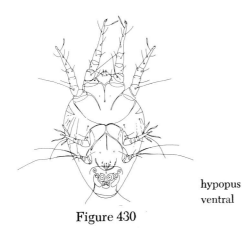

hypopus
ventral

Figure 430

Figure 430 *Vidia cooremani* Baker

Hypopus with solenidion of tarsus I equal in size throughout, not strongly thickened; ventral suctorial plate broader than long; sternum straight, short, free posteriorly; apodemes III and IV united medially; solenidion I rodlike, solenidion II clublike; single lanceolate setae on tarsus I and II, four lanceolate setae on tarsus III, absent on tarsus IV; ventral body setae short, those anterior to suctorial plate very short.

Collected in association with nest of *Ectemnius paucimaculatus* from Maryland.

3 Tarsus IV widens toward its tip; with well-developed gnathosoma; eyes present .... **Genus *Calvolia***

♀ dorsal

Figure 431

♀ ventral

Figure 432

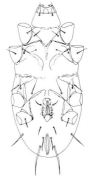

♂ ventral

Figure 433

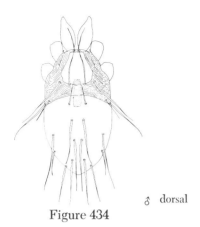

♂ dorsal

Figure 434

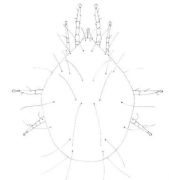
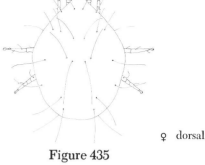

♀ dorsal

Figure 435

Figures 431, 432, 433, and 434 *Calvolia furnissi* Woodring

Adult female with distal-ventral setae of tarsi modified into heavy curved, clawlike structures; propodosomal shield distinct, as wide as long; epigynium large, spermathecae small, with short straight tube; external scapular setae more than twice the length of humeral setae, longer than body setae; legs thin, terminal claw large; male with tarsal IV suckers lacking; aedeagus covered by lateral flaps, long, curved; hypopus with dorsal setae very small; apodemes of III merging with apodemes IV at midline, median epimere absent; three leaflike setae arising from distal portion of tarsus I, II and III, a sickle shaped seta arising from ventral aspect of tarsus I.

Collected in association with galleries and frass of *Dendroctonus pseudotsugae* in Douglas fir from Idaho.

4   Coxal apodemes IV do not meet mesally; metasternum IV absent; internal gnathosomal setae long; tarsi III and IV without digitiform process; claws of tarsi I to III small; apex of tarsus IV pointed, narrows toward its tip ....
.... **Genus** *Monobiacarus*

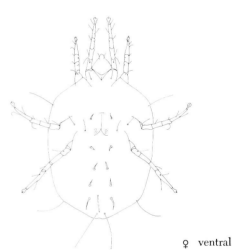

♀ ventral

Figure 436

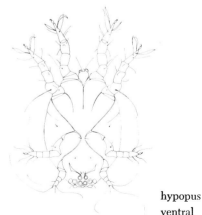

hypopus
ventral

Figure 437

Figures 435, 436, 437 *Monobiacarus quadridens* Baker and Cunliffe

Hypopus with eye separated by a distance subequal to the width of each; tarsus IV with three long whiplike setae, one being abruptly attenuate near its base, one short spinelike, one swordlike, all of medium length; tibia IV with spinelike seta of medium length; tarsus III with empodial claw with inner basal protuberance, with four large lanceolate setae; tibia III with tactile seta and sensory seta.

Collected in association with *Monobia quadridens* from Florida and North Carolina.

5   Internal gnathosomal setae short; tarsal IV seta relatively large, broad, lanceolate; tarsi III and IV without digitiform process; claws of tarsi I, II, III small; apex of tarsus IV pointed, narrows toward its tip ....
.... **Genus** *Kennethiella*

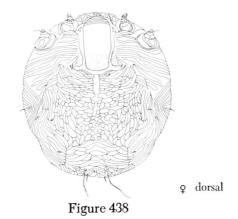

♀ dorsal

Figure 438

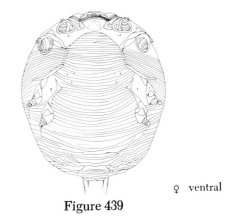

♀ ventral

Figure 439

Figures 438 and 439 *Kennethiella trisetosa* (Cooreman)

Hypopus with anterior propodosomal setae; both pairs of gnathosomal setae short, of equal length; apodemes III indented anteriorly at median junction; sternum reaches past posterior end of apodemes IV; tarsus IV sharp distally, with three long whiplike setae, one short broad lanceolate seta; tibia IV with short, rodlike sensory seta and a tactile seta of medium length; tarsus III with clawlike empodium; female dorsal body setae short; two pairs of genital setae.

Collected in association with *Anoistrocerus antilope* from Washington, D.C.

6   Vertical external setae absent; a division between propodosoma and hysterosoma; supracoxal setae as long as scapular internal setae; genital opening lacking an epigynium; legs without a claw, pretarsus present .... **Genus** *Nanacarus*

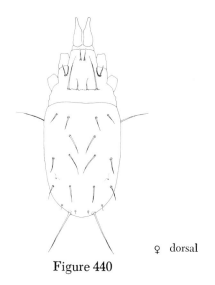

Figure 440

♀ dorsal

Figure 440 *Nanacarus nominis* Woodring

Female without propodosomal shield; external scapular setae and internal scapular setae a microsetae; epigynium absent, genital apparatus very long, almost half length of hysterosoma with a single pair of genital suckers; legs lacking terminal claw, membranous pretarsus present; ventral distal seta enlarged to form claw; sensory setae omega I large and expanded at tip to form a spherical ball.

Collected in association with *Nomia melanderi* from Utah.

Family Psoroptidae

Members of this family have long been known as true mammalian parasites with some of the species considered to be economically important pests of domestic animals. Species of *Psoroptes* have been recorded as causing psoroptic mange to horses, cattle, sheep, goats, elk, rabbit and mountain sheep. Many varieties have been listed for *P. ovis* such as: var. *ovis*, from sheep; var. *caprae*, from goats, var. *bovis*, from cows. The work of Sweatman (1958) is followed in this work. Members of the genus *Chorioptes* cause what is called chorioptic mange to sheep, goats, horses, and rabbits. *Otodectes cynotis* (Hering) is an ear mite associated with dogs, cats and other small carnivores.

1   **Ambulacral stalk long and segmented; cheliceral digits long and narrow (fig. 441)** .... **Genus *Psoroptes***

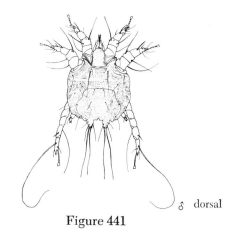

♂ dorsal

Figure 441

Figure 441 *Psoroptes ovis*

The members of this genus are extremely difficult to separate and a worker can only accomplish this by working with a series of individuals which are recognized on a population basis when both biological and morphological criteria are known. Sweatman (1958) constructed a key based entirely on the measurement of the outer opisthosomal seta and stated the following: "It is imperative when a measurement is made that the setae must be measured to its apex and that no broken seta utilized."

KEY TO THE SPECIES OF PSOROPTES (ADULT MALES)

aa   Ear mite ....b
      Body mite c
bb   Adult male with outer opisthosomal seta 64-164 microns ....
      .... *P. cuniculi*
      Adult male with outer opisthosomal seta 145-354 microns ....
      .... *P. cervinus*
cc   Adult male with outer opisthosomal seta spatulate ....
      .... *P. natalensis*
      Adult male with outer opisthosomal sea threadlike .... d
dd   Adult male with outer opisthosomal seta about 333 microns ....
      .... *P. equi*
      Adult male with outer opisthosomal seta 74-258 microns ....
      .... *P. ovis*
°   If adult male with outer opisthosomal seta 145-354 microns and collected in United States ....
      .... *P. cervinus*

Host and Distribution of Psoroptes from Sweatman 1958

P. cuniculi (Delafond) — Ear Mite — Cosmopolitan
    Rabbit, goat, sheep, horse, donkey, mule (also a temporary body mite of horses without occurring in the ears)

P. cervinus Ward — Ear and Body Mite — Western United States
    Bighorn sheep (ear), and Wapiti (body)

P. equi (Hering) — Body Mite — England (only sure country)

P. ovis (Hering) — Body mite — Cosmopolitan
    Domestic sheep, bighorn, cattle, horse, donkey (?), mule (?)

P. natalensis Hirst—Body mite—South America, South Africa, New Zealand — Domestic cattle, zebra, Indian water buffalo, horse

2   **Female perianal region with a pair of setae; legs I, II, and IV with ambulacar, tarsus IV with long terminal setae; coxal apodemes I short, widely separated from apodemes II, ambulacral stalk short, non-segmented; cheliceral digits short and thick; male tarsus III with one long seta; coxal apodemes I short, widely separated from apodemes II; tarsi IV with ambulacra; opisthosomal lobes longer than wide, contiguous at bases .... Genus *Chorioptes***

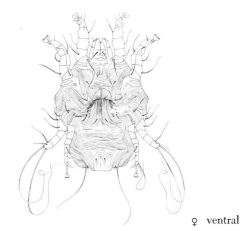

♀ ventral

Figure 442

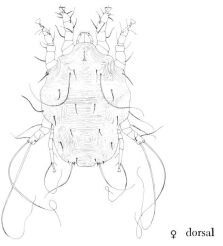

♀ dorsal

Figure 443

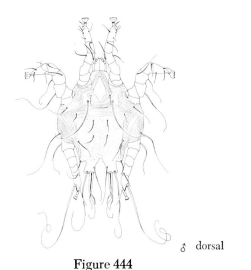

♂ dorsal

Figure 444

♂ ventral

Figure 445

Figures 442, 443, 444, and 445   *Chorioptes bovis* (Hering)

Propodosomal vertical setae and dorsal plates present; female genital opening transverse; sclerotized apodemes present; anal opening posterior, ventral; tarsi I and II clawlike, ending in short-stalked flaplike pretarsi, leg III not strongly developed, ends in a pair of long whiplike setae, leg IV slender, longer than leg III, tarsus ends in flaplike pretarsus; male bilobed posteriorly, long setae arising from each lobe; adanal suckers present on male; leg IV strongly reduced, possesses a short pretarsus.

This is the mite that causes chorioptic mange on animals. This mange is normally localized and not as serious as psoroptic scab or sarcoptic mange. It will be encountered throughout the United States.

3   **Female and male tarsus III with two long, strong setae; coxal apodemes I long, contiguous or fused with apodemes II; female legs I and II with ambulacra, tarsi III and IV with long terminal setae .... Genus *Otodectes***

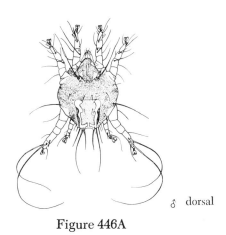

δ dorsal

Figure 446A

Figure 446A *Otodectes cynotis* (Hering)

Genital aperture transverse in female; without vertical propodosomal setae; with sclerotized genital apodemes; tarsi I and II claw-like with short-stalked, flaplike pretarsi; trochanters with no setae; femurs I and II with a single seta each; genu II with two setae and genu I with two seta and a short rod; leg III with long seta on subdistal article; male copulatory suckers present; associated with a pair of short setae near anus; aedeagus between legs II and III associated with a pair of minute setae.

This is the well-known ear mite of cats and dogs. It has also been recorded from ferrets and foxes. The ear mite can be expected to be found throughout the United States associated with dogs and cats.

Family Glycyphagidae

Fain and Whitaker (1973) published on the family Glycyphagidae and followed Fain's (1969) classification of including the Labidophoridae as a subfamily of Glycyphagidae. This classification, Fain (1969), is followed in this work.

The family Glycyphagidae contains members that are associated with stored grain products, nests of mammals, birds, insects, rodents, free-living out of doors, such as the mulch layer of the soil, hay, straw, and can be transferred via air currents from one infested region to another. The species *Glycyphagus domesticus* (DeGeer) is the causative agent called "grocer's itch" of food handlers, *G. domesticus* and *G. destructor* are destructive to cheese. Members of this family are common barn inhabitants of hay lofts, stables, accumulation of all types of crop remains such as corn, piles of straw, stubble, haystacks are also utilized as a place to live. Stores of dried fruit and vegetables, serve as a source of infestation of *Glycyphagus* sp.

The difficult task for a worker when collecting these mites is that they are a part of a complex which includes the mites of the families Acaridae and Carpoglyphidae which are similar and must be separated from Glycyphagid species. Another problem encountered by a worker is that members of this family produce what are known as phoretic hypopi which, unlike the adults and larvae, are associated with animals in a phoretic relationship, not as parasites.

The hypopi are non-feeding forms adapted to cling to other creatures, such as mammals, and their function must be one of distribution of the species. The family Glycyphagidae is not the only family that produces hypopi. The Acaridae, Anoetoidea, Hypoderidae, to name only a few, have certain species that during part of their life cycle produce a hypopal stage.

Some species are only known from the description of the hypopal stage, the adults having never been collected.

Treatment of Adults of the Family Glycyphagidae

1   Vertical internal setae (ve) well separated from vertical external setae (vi); skin rough, granulated; most body and leg setae densely pectinate; tarsi long, slender, small claws attached to long pretarsi; male genital opening between coxae III .... Genus *Glycyphagus*

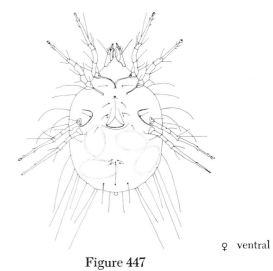

♀ ventral

Figure 447

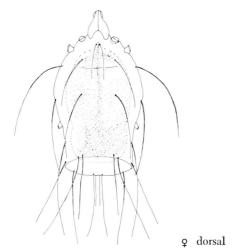

♀ dorsal

Figure 446B

Figures 446B and 447   *Glycyphagus domesticus* (DeGeer)

Supracoxal seta forked, branched; legs long, segments tapering, sub-tarsal scale absent; female genital opening extends to posterior edge of apodemes of legs III, posterior pair of genital setae inserted behind posterior end of genital opening; two pairs of setae on either side of anterior end of anus; tubular bursa copulatrix projects from posterior of body; supracoxal seta much branched, forked rod lying above leg I; genital suckers small; male tarsi I and II not curved; ventral setae of tibiae I and II normal; posterior genital setae long.

This is a cosmopolitan species and is distributed throughout the world and the United States. It has been reported by Krantz (1961) as a part of the succession of granary mites of the Pacific Northwest. Hilsenhoff and Dicke (1963) listed *G. domesticus* as a common cheese mite. Baker et al. (1956) lists this species as one that causes dematitis and is known medically as the mite that causes grocer's itch of humans.

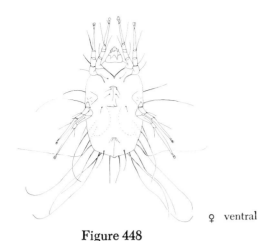

♀ ventral

Figure 448

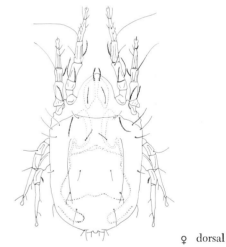

♀ dorsal

Figure 450

♀ dorsal

Figure 449

Figures 448 and 449   *Glycyphagus destructor* (Schrank)

Subtarsal scale present; without a narrow crista metopica; supracoxal seta branched, not forked at apex like *G. domesticus;* two pairs of humerals present; apodemes of legs I meet to form a short sternum, apodemes II well developed, apodemes III and IV much reduced; female genital folds meet, a crescentic epigynium covers anterior fold, bursa copulatrix short, tube with lobed edge.

This species, like *G. domesticus,* is a cosmopolitan species found throughout the world and the United States. This is one of the most common species associated with stored products. It has been reported in all types of cereals, linseed, dried fruit, hay, cheese, straw, corn, insect collections, dried mammal skins, to list only a few of its infestation sites.

2   Setae of dorsal surface short; rostrum without longitudinal ridges; sternum absent; epimeres I joined directly to epigynium; idiosoma sclerotized, brown; bases of leg segments with distinct longitudinal ridges .... Genus *Gohieria*

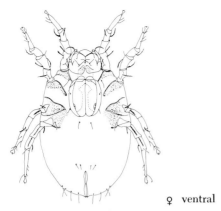

♀ ventral

Figure 451

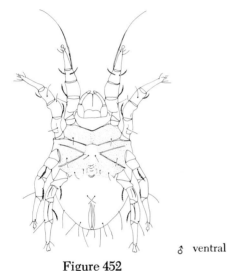

♂ ventral

Figure 452

Figures 450, 451, and 452   *Gohieria fusca* (Ouds.)

Gnathosoma covered by extension of propodosoma, surface of body sclerotized, brown in color, ornamented with fine smooth setae, propodosomal shield very indistinct, crista metopica absent, apodemes of legs slender, unite to surround genital opening, genu and tibiae of legs distinctly ridged, distal edges of femora and genu expanded, anterior edge of hysterosoma marked by transverse ridge, genital opening between coxae IV, aedeagus

straight tube whose apex directed posteriorly, anal opening extends to posterior end of body, one pair of anal setae arises at anterior end.

Collected from mite-infested grain collected in Oregon, Washington and Idaho granaries by Krantz (1961). Hughes (1961) reported that *G. fusca* is frequently found in large numbers in flour, rice, corn bran and pollards in England.

Remarks: Krantz (1961) reported a member of the genus *Ctenoglyphus* as having been collected from mite-infested grain collected in Oregon, Washington and Idaho.

Treatment of Hypopial Stage of Family Glycyphagidae from North American Mammals

The inclusion of the hypopi herein treated is taken exclusively from the works of Fain and Whitaker (1973), Fain (1969) and Rupes and Whitaker (1968).

1   Femur IV with the presence of a recurved apothesis on anterior margin; coxal shield III open, coxal fields IV closed, forming a ringlike coxal area .... **Genus** *Labidophorus*

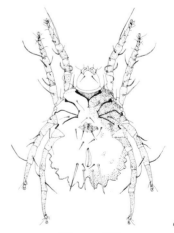

♂ ventral

Figure 453

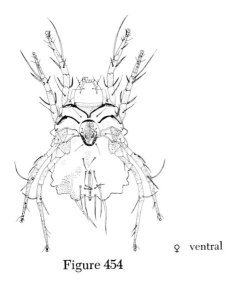

♀ ventral

Figure 454

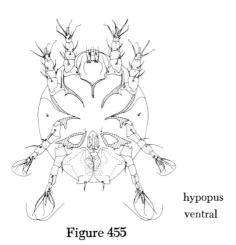

hypopus
ventral

Figure 455

Figures 453, 454, and 455 *Labidophorus talpae* Kramer

Coxal area of legs IV completely enclosed by epimera and epimerites IV, tarsi IV with reduced claws smaller than claw of tarsi III.

Collected on *Parascalops breweri*, hairy-tailed mole, from New York.

2 **Hysterosoma with sclerotized hooklike processes that are recurved; without recurved apothesis on posterior femur;**

epimeres IV open; claw III much shorter than claw I .... **Genus *Orycteroxenus***

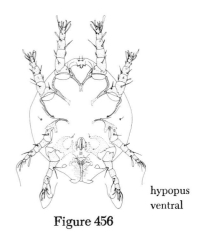

hypopus
ventral

Figure 456

Figure 456 *Orycteroxenus soricis* (Ouds.)

Epimera I separate or contiguous but not fused; trochanters III and IV with projections; absence of triangular cuticular projections laterally to coxa IV; palposomal hair unequal.

Collected on *Blarina brevicauda, Cryptotis parva, Microtus pennsylvanicus, Microtus pinetorum, Sorex arcticus, Sorex longirostris, Sorex palustris* from Rhode Island, Indiana and Minnesota.

Orycteroxenus soricis ohioensis Fain

Epimera I fused into a "Y" with a well-formed sternum; projections absent on trochanters III and IV, presence of a triangular cuticular process on lateral parts of coxa IV.

Collected on *Sorex cinereus* from Indiana, Minnesota and Ohio.

3 **External vertical setae (ve) present; palposoma with two pairs of hairs, one pair of solenidia; claws of tarsi III, IV much smaller and less curved than claws I and II; epimera and epimerites IV converge; clasping apparatus with posterior extremities shaped into triangular**

hooks directed laterally ....
.... Genus *Scalopacarus*

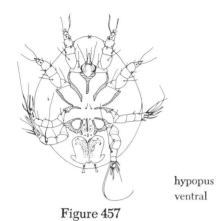

hypopus
ventral

Figure 457

Figure 457 *Scalopacarus obesus* Fain and
Whitaker

Tarsi I and II with only two narrowly foliate
hairs, tarsus IV with a very strong and very
long hair, epimera I fused in a "Y," with the
characters of the genus.

Collected on *Scalopus aquaticus* from
Indiana.

4   Claws I subequal, longer than claws III,
claw IV smaller than claws III; epimera
and epimerites IV either close together
or widely separate; palposoma with vestigial solenidia .... Genus *Marsupialichus*

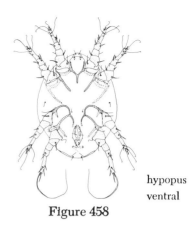

hypopus
ventral

Figure 458

Figure 458   *Marsupialichus johnstoni* Fain

Epimera and epimerites IV widely separate,
coxal region III completely enclosed by fusion
of epimera and epimerites III; claw IV shorter than claw III, dorsum not pitted.

Collected on *Dasypus novemcinctus*
from Texas.

5   Coxal fields III closed; setae ve present;
two pairs of setae and one pair of solenidia on palposoma; epimera I fused in
a "Y," epimera III and IV convergent;
tarsal claws III and IV equal ....
.... Genus *Xenoryctes*

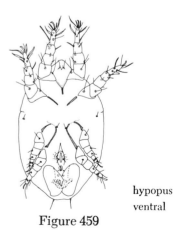

hypopus
ventral

Figure 459

Figure 459   *Xenoryctes latiporus* Fain and Whitaker

Claws I and III equal or subequal, claws III and IV equal or subequal; dorsum pitted, sejugal furrow well developed; epimera III and IV convergent; palposomal solenidia very short; dorsal setae short.

Collected on *Scalopus aquaticus, Spermophilus tridecemlineatus, Mus musculus* from Indiana.

6   Clasping apparatus well-developed, consisting of a pair of large superficial membranous and movable folds and two pairs of deeper ribbed claspers; coxal region III either enclosed by fusion of epimera and epimerites, or epimera and epimerites not meeting; if enclosed, palposoma, with one external pair of setae and one internal pair of well developed solenidia .... Genus *Dermacarus*

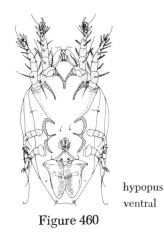

hypopus
ventral

Figure 460

Figure 460   *Dermacarus newyorkensis* Fain

Seta of femur I simple, 80 to 90 microns long, seta of femur II barbed, 35 to 42 microns long; coxal region III not closed, the epimera and epimerites not meeting; claws IV much shorter than claws III; dorsum not pitted.

Collected on *Microtus pennsylvanicus,*

Napaeozapus insignis, Peromyscus maniculatus, Sorex palustris, Zapus hudsonius, Zapus trinotatus from New York, North Carolina, Minnesota, Indiana, Rhode Island and Washington.

7   Pilicolous apparatus normally developed but very small; clasping apparatus on ventral face of opisthosoma very small or completely absent .... Genus *Microlabidopus*

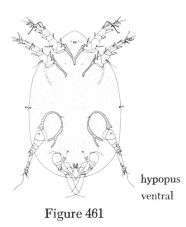

hypopus
ventral

Figure 461

Figure 461   *Microlabidopus americanus* Fain

Pilicolous apparatus normally developed, small; genital suckers situated laterally, close to trochanter IV.

Collected on *Aplodonitia rufa* from Oregon.

8   Pilicolous apparatus absent; clasping apparatus on ventral face of opisthosoma very small or completely absent; genital suckers terminal median .... Genus *Aplodontopus*

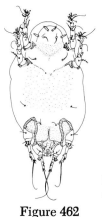

hypopus
ventral

Figure 462

Figure 462 *Aplodontopus sciuricola* Hyland & Fain

Setae vi 17-21 microns long; external palposomal setae 17-20 microns long; pregenital sclerite slightly bifurcate; palposoma at least three times as wide as long.

Collected on *Tamias striatus* from Rhode Island.

Family Acaridae

This family has long been known for its ability to infest stored food products throughout the world. Members of this family are important pests of stored food products in Europe where they have been studied extensively. Despite this economic importance, very little work has been conducted on members found in the United States. The four most important published works dealing with this group are: Zakhvatkin (1941) who studied the fauna of the USSR: Nesbitt (1945) from Canada who revised the family and established that Acaridae should replace the old family name Tyroglyphidae; Robertson (1959) an Australian known for her excellent work on the revision of the genus *Tyrophagus* which is indispensible if attempting to study members of this genus; and Hughes (1961) who authored the book, *The Mites of Stored Food*, based on those members that infest foods primarily in England. A recent work by Manson (1972) of New Zealand treats the genus *Rhizoglyphus*.

Members of this family have been recorded from the Arctic tundra to the tropical rain forests according to Baker and Wharton (1952). These mites have followed man and can be found in all kinds of organic substances. They are commonly found infesting meats, seeds of all types, stored grains, rotting leaves, bark of trees, decaying bulbs, nests of mammals and birds, dog food (both in supermarkets and homes), house plants, cheese, and in some cases they have the ability to increase in numbers to astronomical proportions and can cause economic losses by changing the moisture content and initiating the growth of molds.

Some members of this family produce a hypopial stage similar to that discussed under the family Glycyphagidae.

Tyrophagus putrescentiae, has been reported to cause dermatitis to those persons who come into contact with large populations of this species. Another species, *T. castellanii*, was found infesting the urinary tract and lungs of people and is called the 'copra itch mite.' *Acarus siro* is known as the cheese mite dermatitis or baker's itch mite. According to Baker et al. (1956) dermatitis caused by this group of mites has produced several theories as to the origin of the dermatitis: (1) dust present in the litter created by the feeding mites and viruses; (2) hypersensitivity of the person; and (3) saliva of the mites. Members of the genus *Rhizoglyphus* contain the well-known bulb mite associated with amaryllis, crocus, Easter lily, and gladiola, hyacinth, narcissus and tulip bulbs. *Tyrophagus lintneri*, the widely distributed mushroom mite, is found commonly associated with greasy soda fountains, cupboards, and other food storage areas.

It has been pointed out by Linsley (1944) that the important stored-products pests which are mostly cosmopolitan in their distribution are ignored by most field collectors. Thus their capture is unrecorded, the assumption being that their presence is accidental or resulted from contamination. Because of this, it should be understood that many species of this family known to be present in the United States have not been included in this work as they have not been recorded in the literature.

1   **Propodosomal shield with internal scapulars longer than external scapulars; dorsal setae (d_1, d_2) and lateral (L_a) short in relation to other setae of hysterosoma** .... **Genus *Tyrophagus***

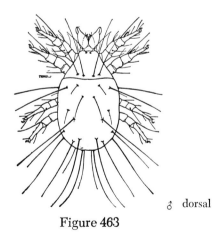

Figure 463   ♂ dorsal

Figure 464

♂ ventral

Figures 463 and 464   *Tyrophagus putrescentiae* (Schrank)

Male with aedeagus S-shaped, aedeagal supporting sclerites turned mediad; supracoxal seta lanceolate, densely set with long setules, attenuate distally, with basal half enlarged, bearing three rows of about five long pectinations; solenidion omega I on first tarsus enlarged at apex; females of this species agree with the male in number of dorsal setae, structure of supracoxal seta arrangement, shape of setae and solenidion; structure of genital opening and arrangement of anal setae not distinctive for separation from other members of the genus *Tyrophagus*.

Collected associated with wheat samples, corn, in *Heliconia mariae*, shops and dwellings, dead larvae of *Anacentrinus subnudus*, Gladiolus corms, stored rice, *Coccolobis uvifera*, corn processing plant, rat cages, straw table mats, citrus leaves, beans, cheese, thrip colony, oranges, orchid plants, pineapple fruit, cucumber leaf, greenhouse, mushrooms, ivy scale, on sprouting potatoes, cockroach cultures, carpets, granary weevil cultures, bee pollen, bacon grease, pet food, elm trunk, cotton seed, pasture mulch, rotten plums, flax seed, rice and wheat from Georgia, Colorado, Alabama, Virginia, Tennessee, North Carolina, Florida, Oregon, Washington,

D.C., Louisiana, Illinois, Maryland, Texas, California, New York, Indiana, Minnesota, South Dakota, North Dakota, Iowa, Delaware, Michigan, Nebraska, Wyoming, New Jersey, Pennsylvania, Maine, New York and Kansas.

This mite will be found in all 50 states. If it has not been recorded it is only because most workers ignore it as the result of contamination.

2   Tarsi dorsodistally with one spine; ventrodistally with three to five spines ....
.... **Genus *Tyroborus***

Tyroborus lini   Oudemans

Appendages usually well sclerotized; supracoxal seta large, gradually widening from apex to base, its edges produced into straight lateral barbs; legs very stout, stalk of claw distinctive, resembling a pair of spines; propodosomal shield pentagonal in shape; ventral coxal apodemes thick, epimeral plates distinct.

Collected associated with mite-infested grain from Oregon, Washington, and Idaho.

Tyroborus lini is regarded as a subgenus of *Tyrophagus* according to Nesbitt (1945), and a synonym of *Tyrophagus lini* by Hughes (1961). Johnston and Bruce (1965) and Samsinak (1962) excluded from *Tyrophagus*, *Tyroborus lini*.

3   Vertical setae "ve" very short, less than one-third the length of chelicera, arise posterior to bases of "vi"; seta d_1 is one and one-half times longer than L_a and d_2 is twice as long as L_a ....
.... **Genus *Forcellinia***

Forcellinia fungivora Oudemans

Supracoxal seta curved, covered with very small pectinations; tarsi I to III with distal seta "e" thickened to form a spine; five ventral spines are present; tarsus IV suckers placed in basal half of segment; aedeagus very long, curved, arising from a basal plate.

Collected in the United States according to Hughes (1961).

4   **Propodosoma with a transverse row of four equally long setae; hysterosoma with one pair of short bristles; tip of tarsus with single dorsal spine; three to five small ventral spines** ....
.... **Genus *Tyrolichus***

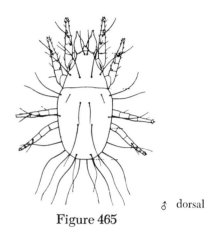

$\mathcal{S}$ dorsal

Figure 465

Figure 465   *Tyrolichus casei* Oudemans

Propodosomal shield square, faintly pitted; vertical setae "ve" more than one-third length of chelicerae; seta L_a is more than twice the length of d_1; legs stout, well sclerotized, ornamented with reticulate pattern, tarsus IV with suckers arising from median portion of tarsus; aedeagus straight, aedeagal supporting sclerites curved inward; supracoxal seta thicken at base with at least nine thin projections; omega I of tarsus I straight, slightly expanded distally.

Collected in association with wheat samples from New York.

Remarks: This is the common cheese mite. Hughes (1961) records this as common in stored food, cheese, grain, flour, old honey combs and insect collections. That it has only been recorded from wheat samples in New York by Krantz (1955) is probably due to

these mites mostly being overlooked or ignored. *Tyrolichus casei* is treated as a member
of the genus *Tyrophagus* by Hughes (1961);
however, Johnston and Bruce (1965), Samsinak (1962) and Zakhvatkin (1941) placed
this species in the genus *Tyrolichus*.

5   **Ventrodistally the tarsi bear one large
spine and two to four smaller setae;
genu of leg one with solenidian omega
I more than three times longer than
omega II, claws of female never bifid;
femur of leg I of male enlarged, bearing
ventrally a conical process, leg I spin-
dle-shaped, much heavier than that of
female** .... **Genus** *Acarus*

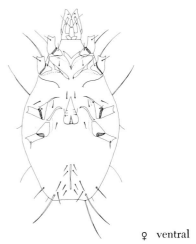

♀   ventral

Figure 467

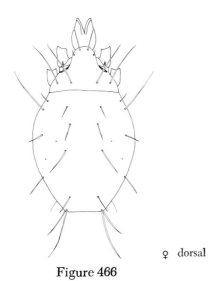

♀   dorsal

Figure 466

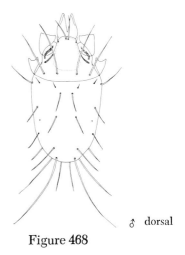

♂   dorsal

Figure 468

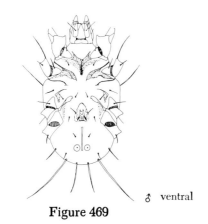

$\male$ ventral

Figure 469

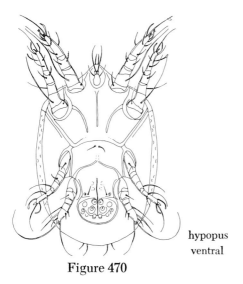

hypopus ventral

Figure 470

Figures 466, 467, 468, 469, and 470 *Acarus siro* Linnaeus

Leg I of male with basal segments enlarged, an apophysis or spur projecting from ventral surface of femur; supracoxal seta expanded basally thickly pectinate; propodosomal shield broad; dorsal setae of hysterosomal region short; seta d_2 less than three times as long as d_1; aedeagus arc-shaped lying beneath genital folds; female genital opening elongated, placed between legs III and IV; legs I no wider than remaining legs, no spur arises from

femur, tarsus IV without suckers, suckers replaced by setae.

Collected in association with mite-infested grain, fungi, wheat samples, mixed feeds from California, New York, Oregon, Washington, Idaho and South Dakota.

This is a common species and is considered to have a distribution ranging throughout the world. There is a hypopal stage. Hughes (1961) records the species as occurring in all kinds of farm related products, hay, barley, flour, cheese, grain, etc. It should be expected to be found throughout the United States associated with food and food-related products, mills, granaries, and seeds.

6 Ventrodistally tarsi bear three large conspicuous spines, inner spine largest; second dorsal terminal seta reduced to a spine, most dorsal setae short; leg I of male scarcely larger than leg I of female, without large spine on femur .... **Genus** *Aleuroglyphus*

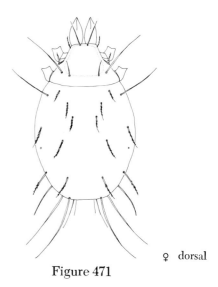

$\female$ dorsal

Figure 471

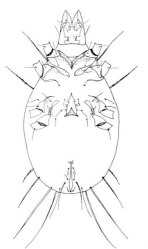

♀ ventral

Figure 472

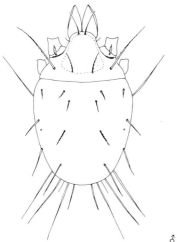

♂ dorsal

Figure 473

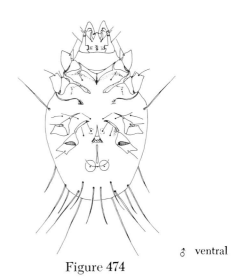

♂ ventral

Figure 474

Figures 471, 472, 473, and 474 *Aleuroglyphus ovatus* (Troupeau)

Chelicerae and legs intensely tanned, dark reddish-brown; setae "ve" long, pectinate; supracoxal seta leaflike, lateral edges produced into straight pectination; tarsus IV with two suckers arising in median part of the tarsus; aedeagus straight, diverging posteriorly; postanal setae are ventral in position, three pairs arranged in the same straight line; internal scapular setae "sci" shorter than external scapular setae "sce."

Collected in association with mite infested grain in Oregon, Washington and Idaho.

7 Dorsal median setae on tarsus I of normal size; vertical setae "ve" represented by short fine setae arising near middle of lateral edge of propodosomal or may be absent; sub-basal and para sub-basal setae present; anal copulatory suckers without sclerotized ring ....
.... **Genus** *Caloglyphus*

Caloglyphus berlesei (Michael)

Male aedeagus a straight tube, well sclerotized; copulatory suckers of tarsus IV located on distal half of the segment; seta d_1 is short, d_2 is two to three times longer than d_1; omega I of tarsus I slightly expanded at distal end; propodosomal dorsal shield slightly concave; supracoxal seta slightly pectinate; epimeral plates well developed.

Collected in association with soil under *Pinus contorta* and mite-infested grain from Oregon, Washington, Idaho, Kentucky and Colorado.

8　Tarsus setae "ba" of legs I and II enlarged to form a stout conical spine; hind portion of propodosoma with a transverse row of four setae of which the inner pairs are minute; male with suckers on tarsus IV, distal sucker almost on rudimentary caruncle ....
.... **Genus *Rhizoglyphus***

Figure 6　*Rhizoglyphus echinopus* Fum. and Rob.

Internal scapular setae "sci" measures less than 5 percent of the idiosoma in length; supracoxal seta bristlelike, longer than dorsal setae one (d_1); aedeagus short, placed between coxae IV; legs with reduced pertarsus; tarsus IV with a pair of suckers placed on distal half; female genital opening between legs III and IV; anal opening surrounded by four pairs of anal setae, external posterior pair much longer than other three; opening of bursa copulatrix terminal.

Collected in association with mite-infested grain in Oregon, Washington and Idaho. This is a common species and will be found throughout the United States.

9　Terminal plate of male opisthosoma divided in a fanlike manner into four distinct lobes; tarsal setae inclined to be spiniform; supracoxal seta minute, body globular-shaped ...... **Genus *Histiogaster***

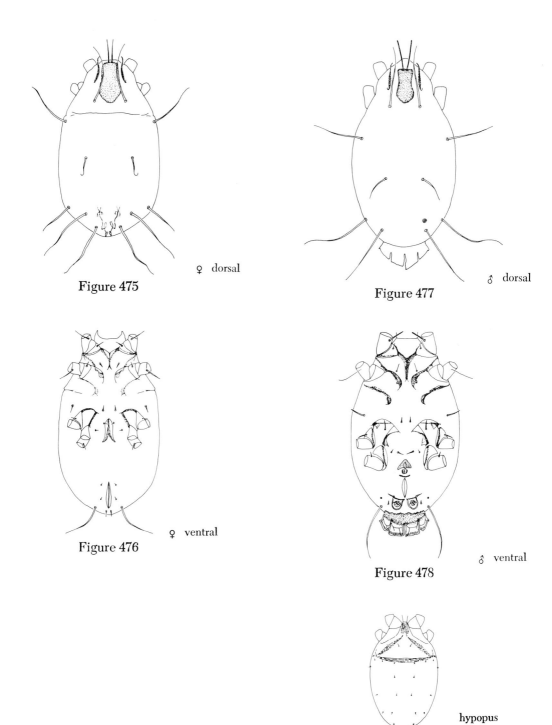

Figure 475
♀ dorsal

Figure 477
♂ dorsal

Figure 476
♀ ventral

Figure 478
♂ ventral

Figure 479
hypopus
dorsal

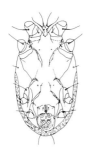

hypopus
ventral

Figure 480

Figures 475, 476, 477, 478, 479, and 480   *Histiogaster arborsignis* Woodring

Female epimeres III and IV fused medially; dorsal seta three not reaching end of body; duct of spermatheca very thin, long, and looped; male opistosomal shield extends anterior on dorsum to level of genital structures bearing a distinct tree or rootlike marking; anterior edge of aedeagus support sclerite forms a smooth arc; hypopus with eye spot elongate; hypostome not longer than wide; posterior and lateral suckers of ventral plate small; genu I and tibia I with no spines.

Collected in association with loblolly pine infested with *Ips* beetle, elm bark beetle, *Pinus echinata* infested with *Ips calligraphus, Ips avulsus, Ips grandicollis, Pinus radiak* infested with *Ips platygaster, I. radiak, Pomphilius luteicornus,* galleries of *D. ponderosae* in *Pinus ponderosa, Ips pini,* from Texas, Louisiana, Delaware, Ohio, Georgia, North Carolina, California, Michigan, South Dakota and Colorado.

10   Hind part of propodosoma with a transverse row of two setae only; terminal setae of tarsus unmodified; male without suckers on tarsus IV and with an opisthosomal shield .... Genus *Schwiebea*

♀ dorsal

Figure 481

♀ ventral

Figure 482

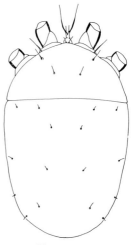

hypopus
dorsal

Figure 483

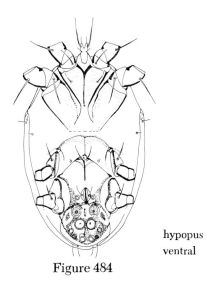

Figure 484

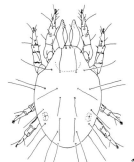

♂ dorsal

Figure 485

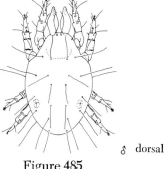

♂ ventral

Figure 486

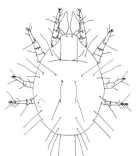

♀ dorsal

Figure 487

hypopus
ventral

Figures 481, 482, 483, 484 *Schwiebea falticis* Woodring

Female propodosomal shield indistinct medially, but deeply cleft; epimeres III and IV joined medially; supracoxal setae a very small button; spermathecae with two chambers, with tubelike duct looping to exterior; male aedeagus small, shorter than diameter of genital sucker; spine overlaying claw of leg III large and conspicuous, leg III tibial sensory setae whiplike; hypopus legs III and IV much thinner than I and II; anterior epimeres not connected to posterior epimeres; coxae I and III bear single seta; ventral plate large, bearing full compliment of suckers.

Collected in association with galleries of elm bark beetle, *D. frontalis* in short-leaf pine from Ohio and Virginia.

11   Hypopus with strong dorsal body setae; coxae strongly sclerotized; gnathosoma prominent; dorsal setae well developed, slender; ventral plate with several pairs of suckers; adult legs lacking longitudinal ridge; male with straight, simple; aedeagus; both sexes with pectinate pseudostigmatic organ .... **Genus *Lackerbaueria***

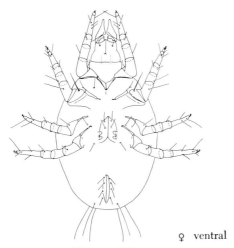

♀ ventral

Figure 488

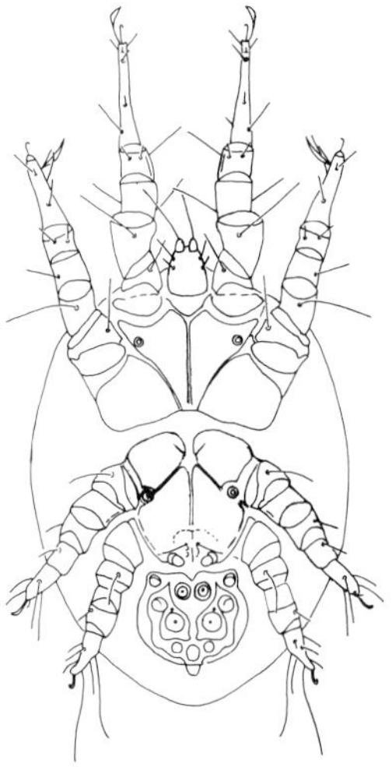

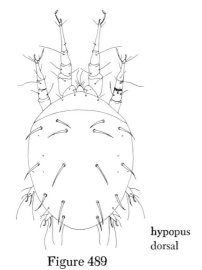

hypopus
dorsal

Figure 489

hypopus
ventral

Figure 490

Figures 485, 486, 487, 488, 489, 490 *Lacker-bauería krombeini* Baker

Hypopus with propodosomal setae short; hysterosomal setae serrate; sternum extending almost to margin of sternal plate; apodemes I and II open, III and IV enclosed; adults with long simple body setae; propodosomal shield well developed; pseudostigmatic organ pectinate; male aedeagus S-shaped, anal discs egg-shaped; posterior anal setae long; tarsal IV discs large, located proximally and distally on tarsus.

Collected in association with adult female of pemphredonine wasp, *Diodontus atratus* and *Passaloecus annulatus* from Virginia.

12   **Hypopus with dorsal reticulate pattern; gnathosoma with two setae bearing tubercles; adults with bifurcate pseudostigmata** .... **Genus *Horstia***

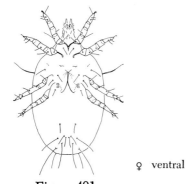

♀ ventral

Figure 491

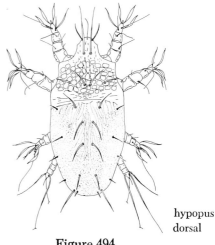

hypopus
dorsal

Figure 494

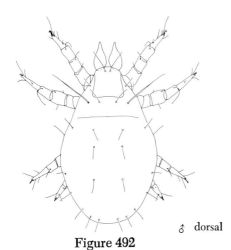

♂ dorsal

Figure 492

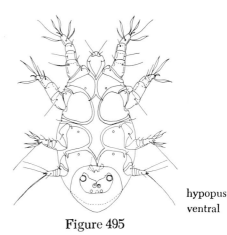

hypopus
ventral

Figure 495

Figures 491, 492, 493, 494, and 495 *Horstia virginica* Baker

Hypopus with propodosoma reticulate; ventral coxal apodemes enclosed, ventral plate with three pairs of large discs; gnathosoma with two tubercles with setae; all tarsi with empodial claws set into tarsi, empodial rods present; tarsus IV with four long whiplike setae; adults with tarsus I sensory rod tapering; male with distinct propodosomal shield; female propodosomal shield indistinct; pseu-

♂ ventral

Figure 493

dostigmatic organ bifurcate; tarsus IV of male with discs located on distal and proximal ends of segment; genitalia of male between coxae IV, anal discs present; genitalia of female situated anterior to coxae III, extending posteriorly to coxae IV.

Collected in association with nest of xylocopid bee, *Xylocopa virginica* from Florida.

13 **Hypopus without gnathosoma; adults with pebbled integument dorsally; male without anal discs .... Genus *Tortonia***

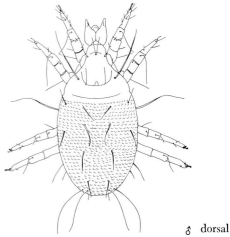

♂ dorsal

Figure 498

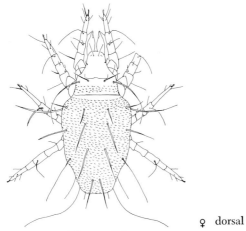

♀ dorsal

Figure 496

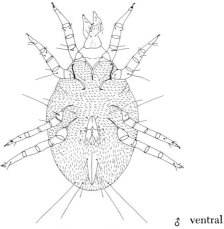

♂ ventral

Figure 499

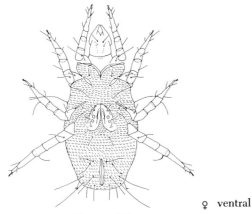

♀ ventral

Figure 497

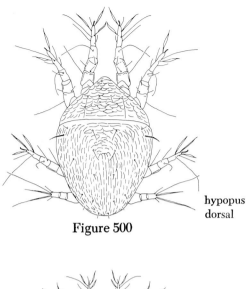

Figure 500

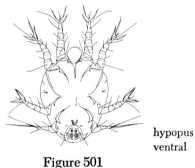

Figure 501

Figures 496, 497, 498, 499, 500 and 501 *Tortonia quadridens* Baker

Hypopus with propodosoma and hysterosoma reticulate; humeral setae of hysterosoma longer than other hysterosomal setae; gnathosoma lacking, only aristae present; ventral apodemes of coxae open, ventral plate with suckers; leg IV without empodial claw, ending in three long whiplike setae; adults with tuberculate integument dorsally; propodosomal shield present; pseudostigmatic organ with six to seven branches; male genitalia between coxae IV; female genitalia between coxae III and IV; anal discs absent in male, discs present on tarsus IV of male, set distally and close together.

Collected in association with vespid wasp, *Monobia quadridens*, from North Carolina.

14 Propodosomal shield absent; idiosoma elongated, longer than wide, distinct transverse groove; sternum well developed; ends of epimeres III and IV joined by chitinous line; coxal fields of III closed .... Genus *Chaetodactylus*

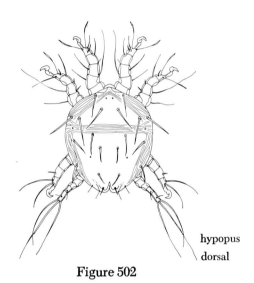

Figure 502

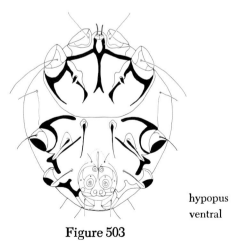

Figure 503

Figures 502 and 503 *Chaetodactylus osmiae* Duj.

Internal scapular setae shorter and thinner than humeral and external scapular setae; internal apical seta of tarsi IV shorter than legs IV, longer than external apical setae; legs I and III of medium thickness, armed with long flexible setae; setae on dorsum short; lateral suckers placed on same transverse line as central ones; posterior suckers equal in size to lateral suckers; legs IV thin, shorter than other legs, ending in one very long macroseta and two shorter setae, internal much shorter than external.

Based on the collection of Banks having reported collecting *Trichotarsus osmia* from a species of *Osmia* from New Jersey.

15   Propodosomal shield well developed, forming two lateral sclerotized bands, which reach base of legs II ending in an oval plate; idiosoma wide, transverse groove absent; male with "V" shaped sternum; female epimeres I joined directly to epigynium, epimeres III and IV short and free .... **Genus *Sennertia***

Sennertia cerambycina Scop.

Posterior edge of hysterosomal shield rounded, strongly reflected on ventral side of body with deep cleft, shield longer than wide; first lateral setae shorter than external scapular setae; external subapical seta of tarsi I and II capilliform; external apical strongly curved, thickened to resemble a knife; ventral seta of tarsi IV small, shorter than tarsus.

Family Anoetidae

The anoetids have two distinct stages within their life cycle: the adults that feed normally where there is an abundance of organic material in a state of putrefaction and high moisture content; and the deutonymphs (hypopi) that are sometimes called the "traveling" or "wander" nymphs which do not feed and may be found in a wide variety of habitats and in association with other animals. Hughes and Jackson (1958) reviewed the Anoetidae pointing out that the life cycle of anoetids may follow two separate sequence stages: the egg-larva-protonymph-trionymph and adult which normally occurs under optimum conditions; or the egg-larva-protonymph-deutonymph-tritonymph and adult. The ability of the hypopial stage to delay the onset of metamorphosis from as short a time as a matter of hours to as long as several weeks depends, it is believed, on the time required for this stage to reach suitable environmental conditions that will provide a source of food for the remaining tritonymph and adults and a median for egg development.

Members of this family have been reported to secrete a substance that will kill fish. Two species are known to be parasitic on earthworms and leeches.

Workers intending to identify members of this family must be aware that many species have only been described from the deutonymphal stage. Therefore, a careful search of the literature must be made to determine if the adults have been described without knowledge of the hypopial form, or the collecting of the adults and not collecting the hypopi. Another difficulty will be that the male, female and deutonymphs require separate keys to determine species groups. Most genera within the family are based entirely on the deutonymph structures and adults utilized to supplement those species when known. This leaves a fertile field for study as much remains to be learned regarding the adults of many of the members of this family.

Since the deutonymph is a traveling stage it many times is collected in a place not suitable for the development of the adults. Therefore, finding the deutonymph in a particular habitat will not normally indicate where to look for the adult stage.

1   **Deutonymph coxa I with disc one; coxa III with disc two; coxa IV with minute seta mesiad of disc three; suctorial plate with two functional suckers, two large central discs, and four discs posteriorly and laterally** .... **Genus *Histiostoma***

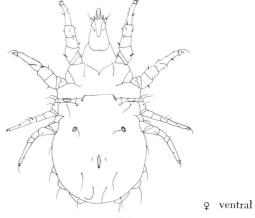

♀ ventral

Figure 505

♂ dorsal

Figure 506

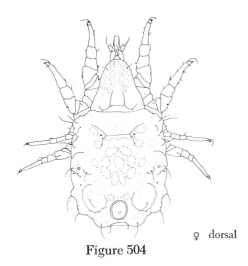

♀ dorsal

Figure 504

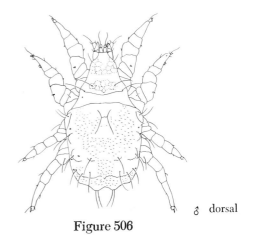

♂ ventral

Figure 507

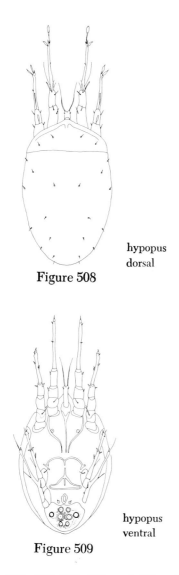

Figure 508

hypopus
dorsal

Figure 509

hypopus
ventral

Figures 504, 505, 506, 507, 508, and 509 *Histiosoma protuberans* Hughes and Jackson

Deutonymph pedipalps short, fused, expanded in mid-region, pedipalpal setae longer than tarsus II; sternum I free posteriorly; apodeme II faintly united with apodeme IV; sternum II short, united with apodeme IV, not connected with apodeme IV anteriorly; large suctorial plate; completely granular dorsal hysterosoma. Adults with chelicera finely serrated bearing approximately 10 fine teeth

distally; anteriorly directed pedipalpal seta I; long ventral setae; unique dorsal evaluations; female bursa copulatrix with surrounding cuplike structure; male with unusual pseudobursa.

Collected in swamp from Virginia and Tennessee.

2   **Deutonymph with minute setae on coxae I and III; coxa IV mesiad of third disc; without eyes; suctorial plate with two functional suckers, two large central discs and four small discs** .... **Genus** *Anoetus*

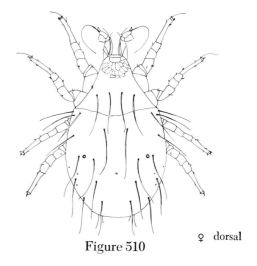

Figure 510

♀ dorsal

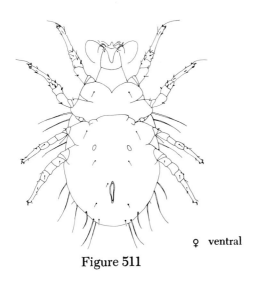

♀ ventral

Figure 511

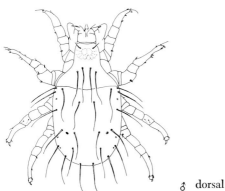

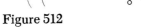

♂ dorsal

Figure 512

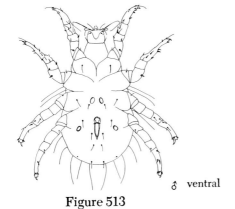

♂ ventral

Figure 513

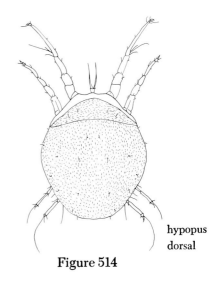

hypopus
dorsal

Figure 514

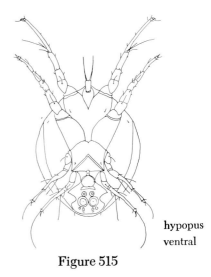

hypopus
ventral

Figure 515

Figures 510, 511, 512, 513, 514, and 515
Anoetus bushlandi Hughes and Jackson

Deutonymph with dorsum shagreen; propodosoma with anterior extension; dorsal setae hairlike; coxal disc III extremely small; laterally indented suctorial plate, plate discs I large; spoon-shaped sensory seta of tarsus IV six times length of claw; adults with finely serrated chelicera; extremely long anterior setae; ringlike structure rounded, small; female bursa copulatrix between opisthosomal dorsal setae one and five.

Collected on stable fly media in Texas.

3   Well divided claw on legs I-III; no spoon-shaped terminal tarsal seta on leg IV; suctorial plate with two functional suckers and two large discs ....
.... **Genus *Myianoetus***

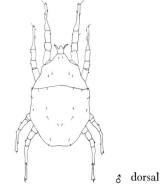

♂ dorsal

Figure 518

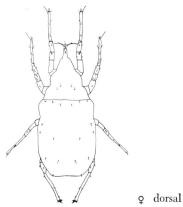

♀ dorsal

Figure 516

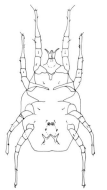

♂ ventral

Figure 519

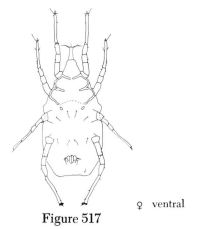

♀ ventral

Figure 517

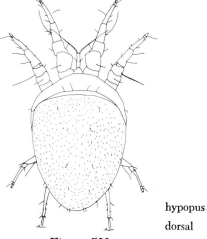

hypopus
dorsal

Figure 520

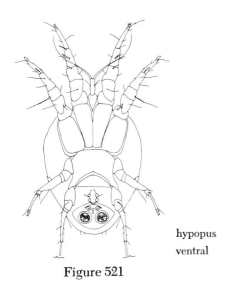

hypopus
ventral

Figure 521

Figures 516, 517, 518, 519, 520, and 521 *Myianoetus muscarum* (Linnaeus)

Deutonymph pedipalpal setae I and II arising nearly from center of fused pedipalps; coxa I and II without discs or setae, coxa IV with setae, without disc; tarsal setae 16 on legs I-II tapering distally; leg I tibia seta III extremely long, displaced to center of tibia; adult chelicera with three equal teeth; conspicuous membranous structures associated with pedipalp; dorsum with scattered pores.

Collected on house flies, *Calliphora terraenovae*, from California and Washington.

Family Pyroglyphidae

The family Pyroglyphidae contains two distinct habitat groups: the sub-family Pyroglyphinae, which is associated with the nests of birds or mammals (Nidicoles); and the free-living members of the genus *Dermatophagoides* which, until recently, were considered to be a part of the families Epidermoptidae and Psoroptidae. This latter group contains the well-known common house dust mite which is of importance since it produces allergic asthma, vasomotor rhinitis, and in some cases, dermatoses in humans. In rare instances, members of the genus *Dermatophagoides* have been recovered from human sputum and urine. They are widespread in their distribution having been recorded throughout the world. Although any of the species within the genus *Dermatophagoides* may be regarded as important from the standpoint of public health, the term, house dust mite is more often associated with two species, *Dermatophagoides farinae* and *D. pteronyssinus*. The work of Van Bronswijk and Sinha (1970) should be utilized as a starting point in working on

members of this family as it is a compilation and evaluation of biologic, ecologic and medical data published between 1964 and 1970. There is included in the latter work a good pictorial key for both males and females for the identification of the pyroglyphid mites of the world.

1 Hysterosomal shield present; external opening of bursa copulatrix and its sperm storage region simple in female; dorsal and lateral setae not located on a shield; external subscapular setae at least five times longer than internal scapular setae.... Genus *Sturnophagoides*

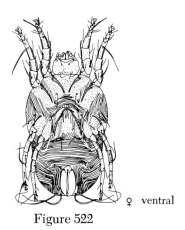

♀ ventral

Figure 522

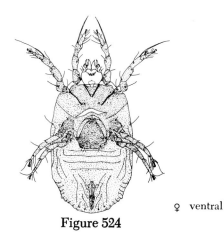

♀ ventral

Figure 524

Figure 522   *Sturnophagoides bakeri* (Fain)

Female with the characters of the genus. There is one other species within the genus, *S. brasiliensis.* The separating feature is the hysterosomal shield which does not reach the posterior apex of the body in *S. bakeri,* and extends to the posterior edge of the body in *S. brasiliensis.*

   Collected from starlings in Virginia.

2   **External subscapular setae the same length as internal subscapular setae; female without bursa copulatrix** ....
.... **Genus *Pyroglyphus***

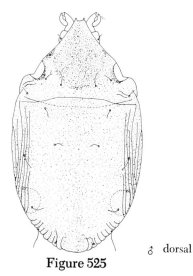

♂ dorsal

Figure 525

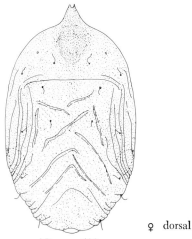

♀ dorsal

Figure 523

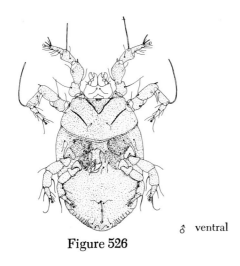

♂ ventral

Figure 526

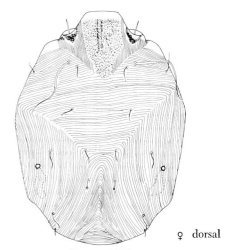

♀ dorsal

Figure 527

Figures 523, 524, 525, and 526 *Pyroglyphus morlani* Cunliffe

Female vulva covered with membrane; male with anterior edge idiosoma with two triangles; aedeagus located between covae III and IV; adults with propodosoma smooth; female genital plate sclerotized with two pairs of setae posteriorly, two small genital discs present, bursa copulatrix absent; tarsal caruncles expanded with small empodial claws, tarsi I and II with single rod-like sensory setae; chelicerae are chelate.

Collected in nest of *Neotoma albigula* from New Mexico.

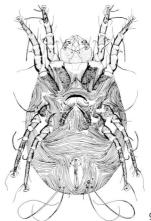

♀ ventral

Figure 528

3   **Hysterosomal shield absent; cuticle with striation simple, without punctation; external scapular setae much longer than internal scapular setae** .... **Genus *Dermatophagoides***

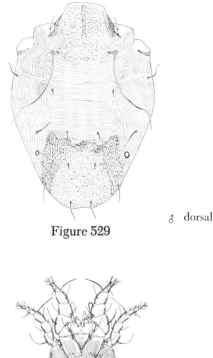

Figure 529
♂ dorsal

Figure 530
♂ ventral

Figures 527, 528, 529, and 530 *Dermatophagoides farinae* Hughes

Female with body striated transversely between dorsal setae two and three; legs and tarsi IV longer than legs of tarsi III; female with external opening of bursa copulatrix ventral near posterior apex of opisthosoma and on a level with a pouch with chitinized walls, internal opening of bursa very narrow and devoid of chitinized structure; male with legs I compressed laterally and much thicker than legs II; apodemes I fused in shape of a "Y"; tarsus III without spines within its medi-

an part; legs III one to five times as long as legs IV.

Recorded from biscuit flour, imported wheat pollards, *Rattus norvegicus, Peromyscus* sp. and scalp dermatitis of humans from Tennessee, North Carolina, Maryland, Massachusetts, and Pennsylvania.

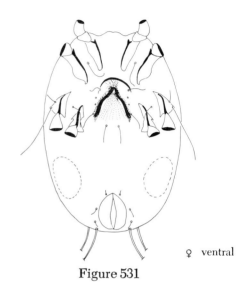

♀ ventral

Figure 531

Figure 531 *Dermatophagoides pteronyssinus* (Trouessart)

Female bursa copulatrix of uniform size, opening behind into a tiny vestibule, in front the spermatotheca has a sclerotized base with the internal apex ending in the shape of a rose or cup; dorsal and lateral setae five, not close together and not within a sclerotized plate region; tarsi III slightly longer than tarsi IV; male hysterosomal plate very short, narrow at its anterior apex; propodosomal plate very narrow in front; tarsus I with one terminal spine, tarsus II lacks terminal spine.

Collected from house dust, overstuffed furniture, birds of the family Tyrannidae and scalp dermatitis in humans from Maryland, Massachusetts, Virginia and Kentucky.

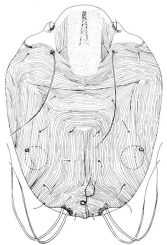

Figure 532

♀ dorsal

Figure 533

♀ ventral

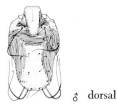

Figure 534

♂ dorsal

♂ ventral

Figure 535

Figures 532, 533, 534, and 535 *Dermatophagoides evansi* Fain, Hughes and Johnston

Female bursa copulatrix enlarged in its posterior wall and opening at its exterior by an orifice in the tube without an intermediate vestibule; spermatheca a sclerified wall and much longer than wide; dorsal and lateral setae five very close together at their bases and in a sclerotized plate region; tarsi III and IV equal; male hysterosomal plate very long, narrow in anterior region; propodosomal plate wide in front; legs III much longer and thicker than legs IV; tarsi I and II respectively terminating in two and one claws.

Collected off *Petrochelidon fulva*, *Quiscalus* from New Mexico and Ohio.

4   Cuticle scleritized, pitted or with striations; legs I and II cylindrical, without membranous chitinized joints, base of leg II without sclerotized pouch; femur II not modified; epistome ending with two points; dorsal and lateral setae five short; genu I with a single solenidion; female legs III shorter than legs IV, genital papille oval; internal opening of bursa copulatrix small, not scleritized .... **Genus** *Euroglyphus*

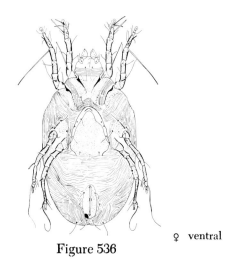

Figure 536   ♀ ventral

Figures 536 and 537   *Euroglyphus longior*
(Trouessart)

Female dorsal scutellum small; cuticle striated with small area pitted, bifid; legs I and II cylindrical without chitinized membranes; genital papille scleritized; male epistome ending with two projections; base of legs II without deep scleritized pouches; dorsal and lateral setae short, lateral five longer than dorsal five; genu I with single solenidion; body narrow at posterior ending with two small lobes; chaetotaxie not reduced.

Collected in dry debris on barn floor from Ohio.

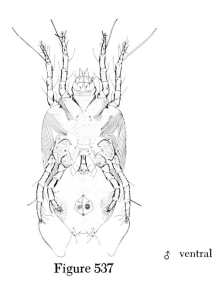

                              ♂ ventral

Figure 537

Family Pneumocoptidae

The mites of this family are found in the respiratory tracts and lungs of rodents. They are very small mites and every care must be taken if these mites are to be collected. The lungs of the host must be macerated in a blender and then digested for 24 hours in potassium hydroxide. The digested mass is then decanted and the sediment searched for specimens.

Anterior and posterior dorsal shields present; only propodosomal sensory setae present; ventrally coxae form large well defined plates; genital opening in both male and female between coxae III; legs with few

setae, each tarsus with long whiplike setae ....
.... **Genus** *Pneumocoptes*

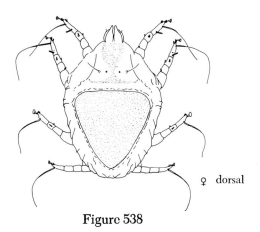

♀ dorsal

Figure 538

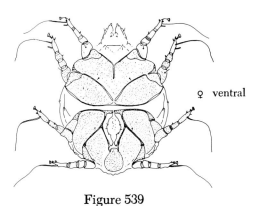

♀ ventral

Figure 539

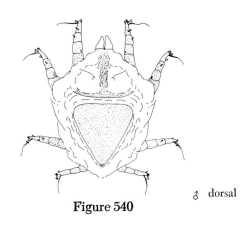

♂ dorsal

Figure 540

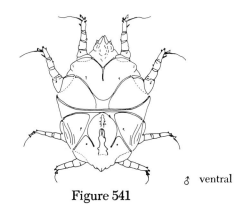

♂ ventral

Figure 541

Figures 538, 539, 540, and 541 *Pneumocoptes jellisoni* Baker

Smaller than *P. penrosei,* female coxal III apodemes straight; area behind genital opening distinctly separated from coxae IV by a suture; male also with straight apodemes of coxal III; a single free genital sucker; aedeagus large, genital area between coxae III forming an angle anteriorly; rostrum projecting anteriorly from body.

Collected in *Peromyscus* sp., *Onychomys leucogaster* from Idaho and Nebraska.

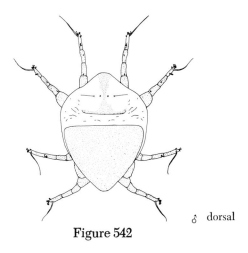

♂ dorsal

Figure 542

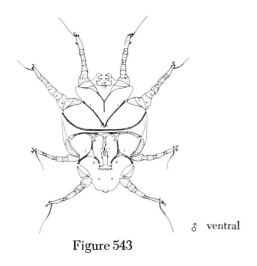

δ ventral

Figure 543

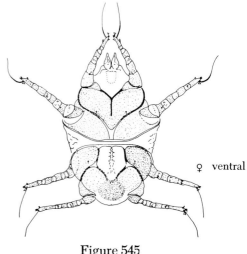

φ ventral

Figure 545

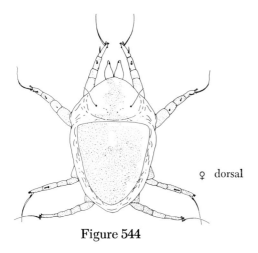

φ dorsal

Figure 544

Figures 542, 543, 544, and 545   *Pneumocoptes penrosei* (Weidman)

Female with inner and posterior apodeme of coxa III forming an angle 130 degrees at junction with coxal IV apodeme; area between coxae IV and behind genital opening poorly defined, not definitely set off from coxal plates; male with three pairs of genital suckers, anterior portion of genital area between coxae III, transverse, not acutely angled; rostrum being covered by body.

Collected on *Cynomys ludovicianus* from Nebraska, Texas and western United States.

Family Listrophoridae

This family, once quite large, was reduced to a more realistic group of species by Fain (1971). They are associated with the fur of mammals and are commonly called fur mites. However, now that a reclassification of the old family Listrophoridae has been established by Mc-Daniel (1968) and Fain (1971), several families, such as Atopomelidae and Chirodiscidae, found in association with the fur mammals can be considered as fur mites. This group is presently undergoing a change in its classification.

1 **Propodosomal shield divided into two dorso-lateral postscapular shields; male hysterosoma with two latero-dorsal**

shields; posterior body margin rounded
with two very small lobes ....
.... Genus *Listrophorus*

♀ dorsal

Figure 546

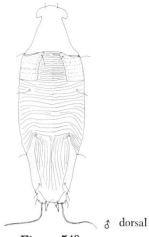

♂ dorsal

Figure 548

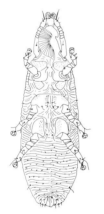

♀ ventral

Figure 547

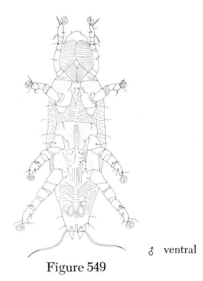

♂ ventral

Figure 549

Figures 546, 547, 548, and 549 *Listrophorus
leuckarti* Pagenstecher

Male aedeagus sickle-shaped, apex curved;
propodosomal shield rhomboidal with single
seta placed at anterior portion; male with
opisthosomal plate divided; anal suckers small,
associated with sclerotized bar; anal region
with single pair of simple setae divided into
two reduced lobes; female larger than male;
opisthosomal region without dorsal plate; gen-

ital area enclosed by apodemes of legs III, two small genital suckers present.

Collected on *Microtus pennsylvanicus, Mus musculus* from West Virginia and Indiana.

2 Propodosoma with two median dorsal shields, one prescapular and one postscapular; coxa II with a sclerotized arch in the shape of an "S"; dorsal setae five (d_5) and lateral setae five (L_5) in female are unequal and have their bases contiguous; legs III and IV thick; aedeagus enclosed in an inverted "U"-shaped sclerite .... **Genus *Lynxacarus***

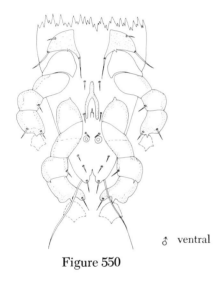

♂ ventral

Figure 550

Figure 550 *Lynxcarus morlani* Radford

Legs III and IV broadly expanded, legs III with three pairs of setae, legs IV with 2 pairs of setae; three pairs of setae on sclerotized region anterior to legs III; aedeagus surrounded by sclerotized bars shaped like a tuning fork, with a pair of setae close to anterior region of sclerotized bar, another pair of setae near base of coxae IV; paired genital suckers posterior to aedeagus with a pair of setae located anterior to suckers; 2 pairs of

setae anterior to anus; a pair of long setae flanking posterior concavity between pair of long terminal setae, dorsal shield covering anterior half of body.

Collected on *Lynx rufus* from Georgia.

3 **External scapular setae dagger-shaped; posterior lobes of male with membranous setae; absence of non-punctated area on the postscapular shields .... Genus *Geomylichus***

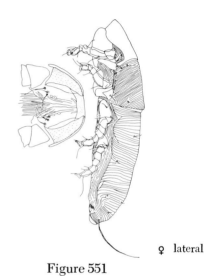

♀ lateral

Figure 551

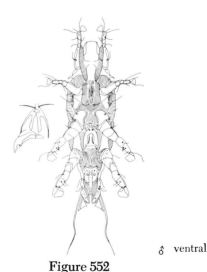

♂ ventral

Figure 552

Figures 551 and 552 *Geomylichus geomydis* Coffman and McDaniel

Male fanlike hyaline membranous setae expanded on inner and outer edges; external scapular setae located on postero-dorsal apices of heavily sclerotized plates immediately posterior to first coxae; opisthosoma divided into two lobes; aedeagus associated between two sclerotized plates; pair of microsetae below apex of aedeagus; female genital opening between legs III; three setae on each side of genital opening; setae arising from antero-dorsal apices of tibiae of legs I and II equal to similar setae on tarsi; male with same setae more than twice as long.

Collected on *Geomys bursarius bursarius, Thomomys umbrinus agricolaris, Geomys personatus megapotamus, Geomys bursarius illinoensis,* from South Dakota, California, Texas and Indiana.

4   Postscapular shield with a medial oval non-punctated zone in medial region; hysterosomal dorsal region with a medial shield or two dorsolateral shields; male with posterior lobes not well-developed and without membranous setae .... Genus *Prolistrophorus*

Prolistrophorus bakeri (Radford)

Caudal region of male with a hyaline flap, without lobes; anal setae without membranous setae; aedeagus short, tapering, associated with sclerotized apodemes, placed between coxae IV; subscapular shield with median correction at posterior margin of shield, internal scapular setae placed on subscapular shield, external scapular placed on head shield; five opisthosomal setae, three on opisthosomal shields; opisthosomal shields wide, covering lateral area of opisthosomal region.

Collected on *Sigmodon hispidus, Oryzomys palustris, Peromyscus gossypinus* from Georgia and South Carolina.

5   Body in both sexes broadly egg-shaped; tergmen very wide and rounded in front; anal suckers with a small medial chitinous axis; male with posterior region elongated to form wide flat appendages that divide apically into two wide terminal lobes each containing a large membranous setae; posterior legs of male slightly expanded ....
.... **Genus *Leporacarus***

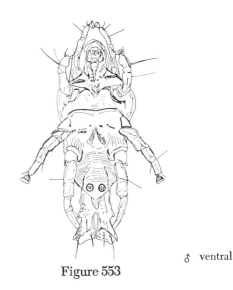

Figure 553   ♂ ventral

Figure 553 *Leporacarus gibbus* (Pagenstecher)

Male with posterior portion of body with caudal prolongation; female without median hysterosomal shield, with complete fusion of propodosomal shields; caudal lobes of male broadly shaped; hysterosomal shield with warts; setae of caudal membranous lobes not striated; aedeagus placed between coxae III and IV; apodemes of coxae III joined to make a transverse bar across venter, apodemes of coxae II connected forming a "Y"-like structure with the arms attached at the center.

Collected on *Lepus californicus, Rattus rattus* for Georgia, Texas and South Dakota.

6 Body egg-shaped, without caudal elongation; dorsal setae five (d_5) and lateral setae five (L_5) of female unequal but base not contiguous; with anglelike annulations on opisthosomal region; postscapular plate present; male with hook- or curve-shaped aedeagus, without associated pregenital sclerites ....
.... **Genus *Olistrophorus***

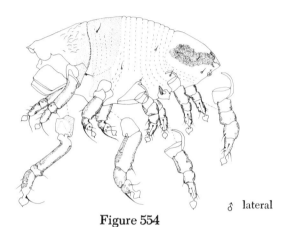

Figure 554

♂ lateral

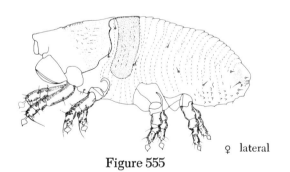

Figure 555

♀ lateral

Figures 554 and 555   *Olistrophorus cryptotae* McDaniel and Whitaker

Male aedeagus broadly expanded with apex extended into a small hook, broad curved region with a sharp pointed spinelike projection; opisthosomal plates not divided into a broad anterior portion, connected by a narrow extension to a broad posterior section; body egg-shaped; female opisthosoma and metapodosoma with two types of annulations, dorsal region with typical linelike annulations, ventrally replaced by small anglelike humps; a pair of microsetae located between coxal apodemes of legs I and II.

Collected on *Cryptotis parva* from Texas.

7 Tegman flat, weakly sclerotized, not covering whole gnathosoma; coxae II widely separated; sternal region very wide covered by chitinous striated membranes extending from epigynium to gnathosoma; palps with conspicuous membranes; legs I modified into attaching organ which is "S" shaped; tibia incurved outward; tarsus and ambulacra curved inward; body strongly flattened ....
.... **Genus *Aplodontochirus***

Aplodontochirus borealis Fain and Hyland

With the characters of the genus; dorsum bears large sclerotized shields; three in the female, four in the male; all legs with large suckers; male with two adanal suckers; posterior extremity divided into two partly membranous lobes; female 336 microns long (gnathosoma included) and 141 microns wide; male 310 microns long and 134 microns wide.

Collected on *Aplodontia r. rufa* from Washington.

Remarks: No figures were included by the authors.

Family Atopomelidae

This family contains mites associated with the fur of mammals that was formerly a part of the family Listrophoridae. Most of the species within this family are found in Africa, Australia and South America. Only two species of this family have been recorded from the United States.

Body dorso-ventrally compressed; gutter pilicole wide; dorsal region with two medial propodosomal shields and one hysterosomal shield; sternum long, entire narrowing behind, bearing well formed adanal suckers; tarsus IV bluntly curved inwardly in the apical portion, terminates by several small triangular points; sucker IV inserted in prolonged axis of the legs .... Genus *Chirodiscoides*

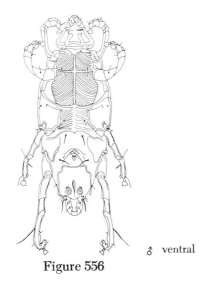

♂ ventral

Figure 556

Figure 556 *Chirodiscoides caviae* Hirst

Sternum not divided into two halves in posterior half; hysterosomal shield of female broad and not rounded posteriorly, many small scales on ventral opisthosoma; male without transverse diamond-shaped shield between epimeres IV; intercellular shield absent;

tarsus IV bluntly and strongly curved inwardly in apical portion; sternum narrowing behind; tegmen absent; anterior legs modified, strongly shortened.

Collected from the domesticated guinea pig throughout the United States. It also is found on white mice in research laboratories.

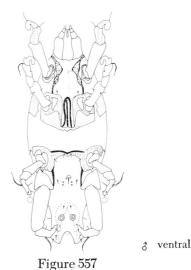

♂ ventral

Figure 557

Figure 557 *Chirodiscoides oryzomys* (Radford)

Male clasping apparatus striate between coxae II as well as coxae I; mid-dorsal shield very narrow medially reaching to level of posterior edge of leg II; hysterosomal shield deeply cleft posteriorly; aedeagus slender, curved; rod-like transverse coxal apodemes between legs III and IV; female postdorsal shield truncate, extending to level of coxae IV and apex of hysterosoma, concave posteriorly; genitalia between coxae III; legs III and IV of normal size; tarsus III with enlarged dorsobasal seta.

Collected on *Oryzomys palustris* from Georgia.

Family Sarcoptidae

The itch mites are well known because of their association with humans. These mites are spread by contact with an infected person, usually by sleeping in the same bed. The work of Mellanby (1943) entitled *Scabies* gives detailed information regarding the spread of scabies in humans. Normally it is the fertilized female mite tunneling just beneath the skin's surface to lay their eggs that causes the itching. The immature stages are associated with the hair follicles or on the skin surface with the male.

The infestation of members of this family on domestic animals is known as sarcoptic mange or scab. Records have shown that sarcoptic mange of domestic cattle and sheep can cause the death of the animal. The genus *Notoedres* has been recorded from a wide range of mammals. They cause what is termed by most workers as notoedric mange. The first species described for the genus *Notoedres* was *N. cati* (Hering) taken from crusts on the head of a mangy cat.

1   **Dorsal striae broken by strong, spine-like serrations; posterior dorsal body setae short, lanceolate** .... **Genus Sarcoptes**

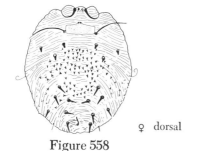

♀ dorsal

Figure 558

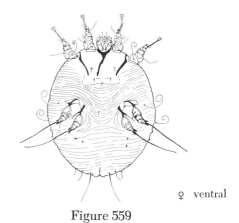

♀ ventral

Figure 559

Figures 558 and 559   *Sarcoptes scabiei* (Degeer)

Vertical setae on anterior median margin of propodosoma; skin striae broken by spinelike projections; dorsal hysterosomal setae stout, lanceolate; legs short, leg I and II end in long stalked flaplike pretarsi, legs III and IV end in a single, whiplike setae; female genital opening a simple transverse slit; male smaller than female; epimeres of legs III and IV fused, united by a transverse bar from the middle of which a chitinous structure in the form of an inverted "Y" proceeds backwards and forms the genital armature; genital structure placed between fourth pair of legs; anus terminal.

Remarks: Mites of the genus *Sarcoptes* found on humans are structurally similar to those attacking domestic animals. However, Baker et al. (1956) considers the forms associated with domestic animals as biological races not easily transferred from one host to another. Sarcoptic mange is associated with domestic animals throughout the United States. The mites cause serious reduction in meat, milk, or wool production and extreme discomfort to humans.

2 Dorsal striae not broken by spinelike serrations; anus dorsal; tarsi with long pretarsi on legs I and II .... **Genus *Notoedres***

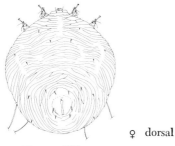

♀ dorsal

Figure 560

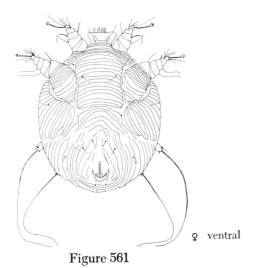

♀ ventral

Figure 561

Figures 560 and 561 *Notoedres muris* (Megnin)

Body subglobular, long, 400 microns, with distinct dorsal anus; scapular setae very short (maximum 12 microns); preanal setae thick, some spear-shaped, maximum length 15-18 microns; pretarsal suckers long, from 14-18 microns; dorsal region between dorsal setae d_1, d_4 and lateral setae l_5 with normal striations; preanal setae very small; dorsal setae simple; anus subterminal; scales entirely lacking on dorsal surface; one setae present on coxae IV; epimeres "Y" shaped; idiosoma globular.

Collected from *Microtus californicus* from California.

Family Epidermoptidae

Members of this family are skin parasites of birds and, according to Fain (1965), some of the members are true hyperparasites. Members of the species *Myialges macdonaldi* (the immatures, males and the non-gravid females) live in the skin of birds and the fertilized females are found fixed on a hippoboscid fly that is also a parasite on the bird host. Fain and Bafort (1963) observed that the female oviposit on the hippoboscid fly. The hatched larvae then return to the bird host on which they feed; the spread of this species is via the larvae. A worker attempting to classify members of this family should consult the works of Fain (1965): *A Review of the family Epidermoptidae Trouessart Parasitic on the Skin of Birds.* The species *Epidermoptes bilobatus* has been collected under the skin pellicule of domestic fowl. *Rivoltasia bifurcata,* according to Krantz (1970) can cause intense itching and scrufiness on the head of chickens.

1   **Tarsi III with less than six setae; female epimera I separated posteriorly by a small epigynium that is contiguous; male with tarsi IV reduced; posterior margin of opisthosoma entire** ....
.... **Genus** *Epidermoptes*

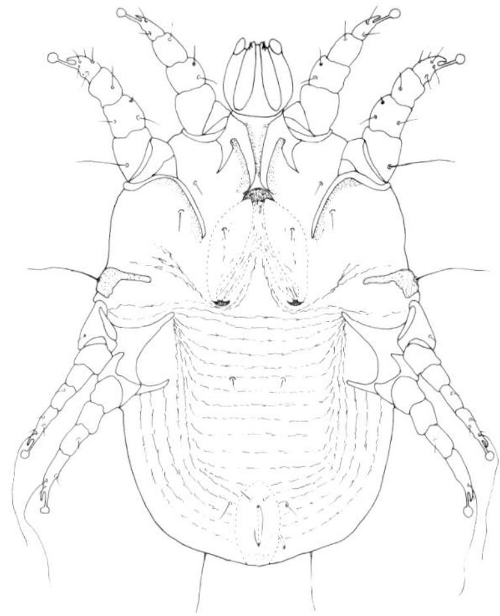

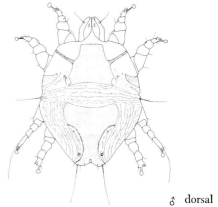

Figure 563

♀   ventral

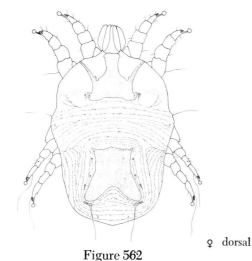

Figure 562

♀   dorsal

Figure 564

♂   dorsal

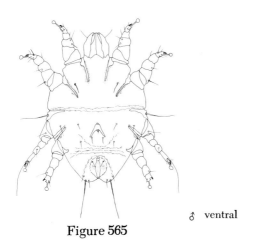

♂ ventral

Figure 565

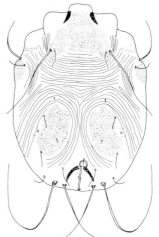

♀ dorsal

Figure 566

Figures 562, 563, 564, and 565 *Epidermoptes phasianus* McDaniel and Parikh

External scapular setae (sc e) short, larger than internal scapulars; female hysterosomal shield with posterior region bilobed; genital lobes an inverse Y with posterior portion of arms bearing sclerotized semicircle projection; bursa copulatrix behind anal opening; male propodosomal shield with two arms projecting laterally to margin of body; genital region with three pairs of setae, one pair near anterior end of aedeagus, two pairs of setae posterior to aedeagus; anal suckers small.

Collected on *Phasianus colchicus*, ringnecked pheasant from South Dakota.

2   Palpal tarsus with a subapical thick and blunt hair recurved outside; tarsi I and II equal with large, strongly recurved clawlike process; trochanter III and IV with a flat ventral process directed internally .... Genus *Microlichus*

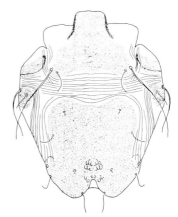

♀ ventral

Figure 567

Figure 568

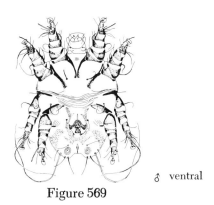

Figure 569

§ ventral

Figures 566, 567, 568, and 569 *Microlichus americanus* Fain

Female hysterosomal shields widely separated, elliptical, each shield 100 to 150 u long and 60-65 u wide; epimera I convergent, not contiguous, epimera II to IV with large wide punctured bands; chitinous crescentic band behind anus not interrupted in middle; dorsal seta of palpal tibia long, thin with conical base; male with short aedeagus, presence of a u-shaped chitinous band surrounding genital organ; tarsus IV modified with two suckerlike seta.

Collected on *Euphagus carolinus, Dumetella carolinensis, Lophortyx californica* from Florida and Ohio.

3   Palpal tarsus without a subapical thick and blunt recurved hair; trochanter III and IV without ventral process; male posterior lobes and adanal suckers lacking .... **Genus Myialges**

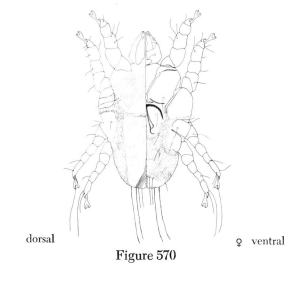

dorsal   ♀ ventral

Figure 570

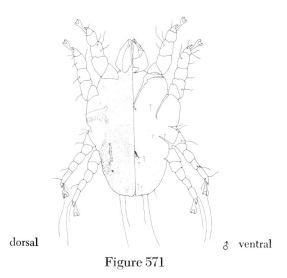

dorsal   ♂ ventral

Figure 571

Figures 570 and 571 *Myialges bubulcus* (McDaniel and Price)

Female endogynial arc long, extending beyond level of inner ends of apodemes of coxae II, with a median toothlike projection on endogynial arc; legs I and II with claws subequal, shorter and thicker than those on legs III and IV; apodemes of coxae I fused two-thirds of distance from anterior ends, not fused to endogynium; auxiliary sclerite placed between apodemes I and II; three setae located posterior of endogynium; four long anal setae; male with auxiliary sclerite absent; anal region with four pairs of setae; aedeagus opening an inverted "V."

Collected on *Bubulcus ibis* from Texas.

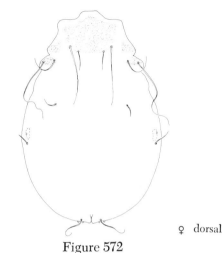

♀ dorsal

Figure 572

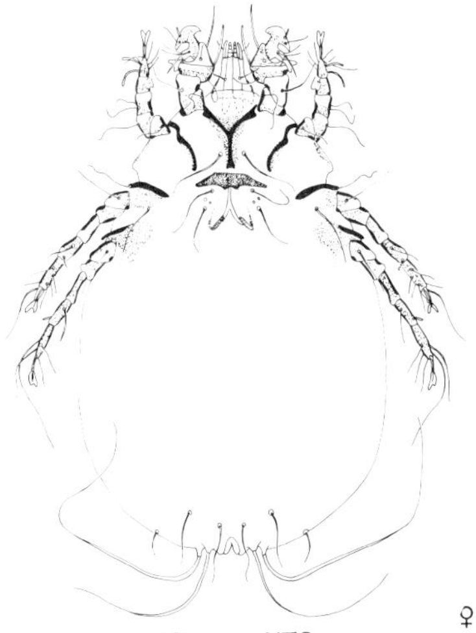

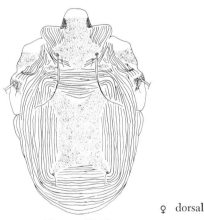

♀ ventral

Figure 573

Figures 572 and 573 *Myialges anchora* Trouessart.

Female with chelicerae long, (87-92 microns); propodosomal shield large (115 microns long; 130-140 microns wide); tibia II with a ventral retrorse process; opisthosoma broken posteriorly; epimera I fused into long sternum; genital apodemes forked, genital setae thin, long; tarsi I with at least seven simple setae, five on base of tarsus, two or three on anchor process; tarsi II with eight setae, five in basal half, three apical; tarsi III with six seta; tarsi IV with five setae.

Collected on *Ornithoica vicina, Lynchia hirsuta, Lophortyx californica vallicola, Lynchia fusca,* taken off *Bubo virginianus.*

4   Legs I and II long, cylindrical, devoid of terminal clawlike process; setae l_4 lacking, setae l_3 present; genu II with only one seta; male with legs IV always much shorter and thinner than legs III .... **Genus** *Dermation*

♀ dorsal

Figure 574A

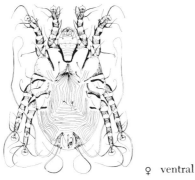

Figure 574B   ♀ ventral

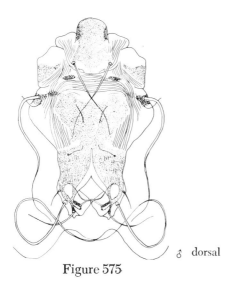

Figure 575   ♂ dorsal

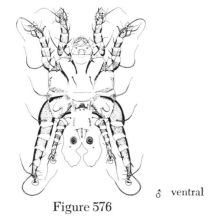

Figure 576   ♂ ventral

Figures 574, 575, and 576   *Dermation anatum* Fain

Propodosomal shield roughly pentagonal; pulvilli with small cylindrical transparent prolongation; female hysterosomal shield narrow in its median part, with posterior angles long and narrow, anterior and posterior margins strongly incised; epimera II, III, IV free, epimera IV forked internally; tarsi I and II with seven setae, tarsi III and IV thinner than their correspondant tibiae and without hooks; male with posterior lobes recurved inside, with well developed membranes; legs III almost twice as long as legs IV, tarsus ending in two recurved clawlike prolongations.

     Collected on *Aix sponsa* from North America.

5   Setae 1_3 and 1_4 lacking, setae sc i shorter and weaker than sc e; solenidion omega I of tarsus I completely absent; legs IV never thinner than legs III, tarsi III and IV with only four setae, strong recurved hooks absent on propodosoma .... Genus *Passeroptes*

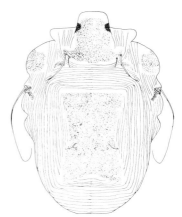

♀ dorsal

Figure 577

♀ ventral

Figure 578

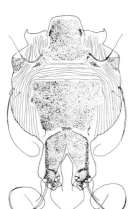

♂ dorsal

Figure 579

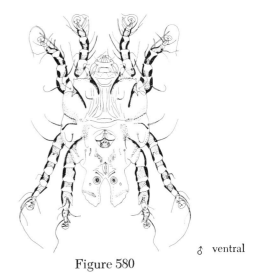

♂ ventral

Figure 580

Figures 577, 578, 579, and 580   *Passeroptes temenuchi* Fain

Epimera III and IV fused by means of a chitinous band; anal shields reduced in their anterior region; solenidion omega vestigial, on tarsus I; hysterosomal shield slightly excavated margins; male with ventral surface of tibiae IV with a chitinous and well sclerotized process ending in a blunt point directed apically; aedeagus short; legs IV shorter but thicker than legs III; femur III with small blunt ventral process.

Collected on *Sturnus vulgaris* from Ohio.

Family Chirodiscidae

This family was established by Fain (1971) on the basis of its priority over Labidocarpidae. The latter was erected from Labidocarpinae Gunther (1942) by McDaniel (1968). The utilization of the family name Chirodiscidae is believed to more closely reflect the proper morphological relationship within the superfamily Listrophoroidea. It will not pose a problem to a worker in using the host specificity that exists for some of the members of this family since subfamily rank has been accorded to the labidocarpids which are parasites of bats. The recent work of McDaniel and Coffman (1970) treated all of the labidocarpid bat mites recorded from the United States.

1 Propodosomal setae minute, barely exceeding length of base ....
.... Genus *Alabidocarpus*

♀ lateral

Figure 581

♂ lateral

Figure 582

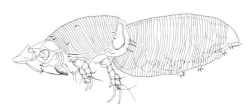

Figure 583

Figures 581, 582, and 583   *Alabidocarpus calcaratus* Lawrence

Both sexes contain a pair of short incrassate lanceolate spines between epimera of leg IV; two pairs of lateral setae in region of coxae III; propodosomal shield well developed, extending to coxae II, without acute posterolateral projections.

Collected on *Myotis yumanensis saturatus* from California.

2 Propodosomal region with a distinct annulated crest in dorsal edge between median narrow rodlike plates ....
.... Genus *Olabidocarpus*

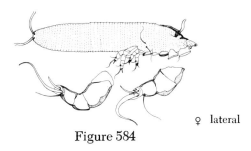

Figure 584

$\female$ lateral

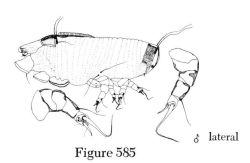

Figure 585

$\male$ lateral

Figure 584 and 585 *Olabidocarpus lawrencei* McDaniel and Coffman

Legs IV with only a single spur associated with main claw; legs III with single main claw, ending in a fine hairlike point, longer and thicker than three accesory spurs, inner accessory spur larger than other two, knife-like, with furrowed curved outer surface, apex pointed; legs IV with main claw longer than main claw of legs III, with one accessory spur; propodosomal plate elongate, with two lateral projections, each beset with a single setae, median propodosomal plates consisting of a pair of long narrow rodlike plates separated from main propodosomal plate with a pair of setae at anterior apex, separated along midline by manelike crest composed of large annulations extending to level of apodemes of legs III.

Collected on *Tadarida brasiliensis mexicana* from Texas.

3 Propodosomal region without a distinct annulated crest; propodosomal plate divided into two parts, anterior part bifid, posterior part with lateral arms bearing swordlike setae ...... Genus *Dentocarpus*

Dentocarpus macrotrichus Dusbabek and Cruz

Propodosomal region without a distinct annulated crest; propodosomal plate divided into two parts, anterior portion bifid, posterior portion with distinct lateral projections, each bearing setae; anterior head plate with horn-like projections; female without knobed posterior region of abdomen; legs III with three accessory spurs and a main claw, legs IV with two accessory spurs and a main claw.

4 Tarsus III and IV without cone-shaped spines, anteriorly, but ends with well-shaped stemmed sucker, body subcylindrical .... Genus *Schizocarpus*

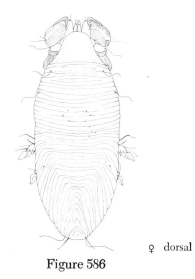

$\female$ dorsal

Figure 586

♀ ventral

Figure 587

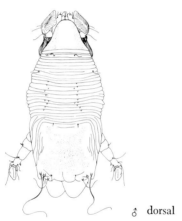

♂ dorsal

Figure 588

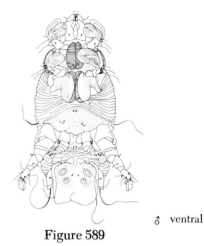

♂ ventral

Figure 589

Figures 586, 587, 588, 589 *Schizocarpus migaudi* Trouessart

Legs III and IV without claws, with circular discs, leg III with stout spine; absent on leg IV; body near leg III with short setae, two other setae placed posterior to head shield; two long anal setae associated with a pair of small setae.

Collected from *Castor canadensis* from California, Michigan, Minnesota, South Dakota, North Dakota, Texas, Washington and Oregon.

Family Myocoptidae

The Myocoptidae are parasites commonly associated with the fur of mammals. *Myocoptes musculinus* (Koch) is the cosmopolitan mange producing species in laboratory mice throughout the world where mice are used as research animals. It is commonly found on the house mouse, *Mus musculus,* which is believed to be its natural host and from which it has spread to other rodents that associate with or frequent buildings, runways or nests of the house mouse.

1   **Anus ventral; epimeres of legs I separate; epigynium absent in females with platelike projections on hysterosoma and venter; tarsus III and IV without fingerlike clamplike seta .... Genus *Myocoptes***

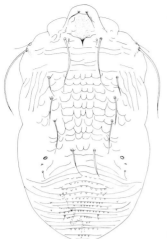

♀ dorsal

Figure 590

♀ ventral

Figure 591

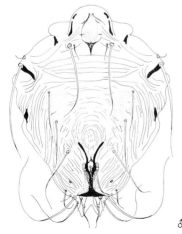

♂ dorsal

Figure 592

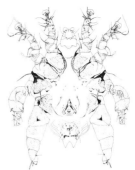

♂ ventral

Figure 593

Figures 590, 591, 592 and 593 *Myocoptes musculinus* (Koch)

Male with ventral adanal shields very small; posterior lobes with a wide angle; inner scapular setae sc i small, outer scapular setae sc e much longer than inner scapular setae; dorsal setae d_1, d_2 similar in size to outer scapular setae sc e; propodosomal plate very lightly sclerotized; hysterosomal region with lightly sclerotized plate shaped as a "U"; genital region sclerotized, elongate; anal setae long; female without sclerotized epigynium, small platelike pattern on venter extending from legs IV to anus; internal scapular setae much smaller than anterior setae.

Collected on *Mus musculus* throughout the United States. Also collected on *Peromyscus leucopus* from Rhode Island.

2   Anus dorsal; female bursa copulatrix dorsal; epimeres I fused in the shape of a "Y" in the female and a "Y" or "V" in the male, tarsus III and IV with finger-like setae .... Genus *Trichoecius*

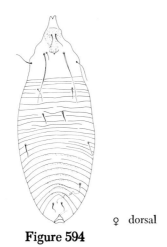

♀ dorsal

Figure 594

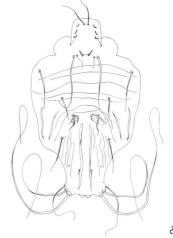

♂ dorsal

Figure 596

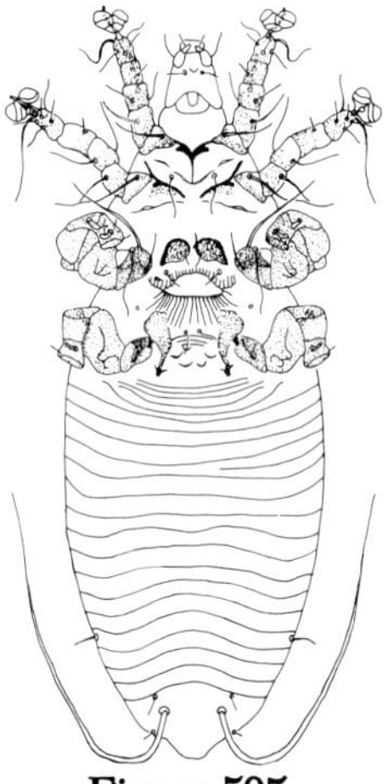

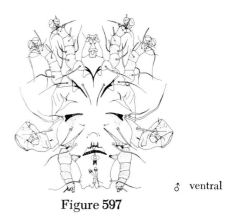

♀ ventral

Figure 595

♂ ventral

Figure 597

Figures 594, 595, 596, and 597 *Trichoecius tenax* (Michael)

Male with ventral setae below aedeagus not bifid; opisthosoma elongated, narrowing from region of coxa IV, legs IV approximately as long as opisthosomal region; aedeagus located between legs IV, setae associated with coxae IV consisting of a very short setae next to center of venter, and a setae longer and extending beyond body region; female dorsal setae d_1, d_2 equal in size and shape.

Collected on *Microtus pennsylvanicus* from Rhode Island.

3 Epigynium well developed in female; male with pointed spinelike projection on legs I and II, setae d₃ long .... **Genus** *Sciurocoptes*

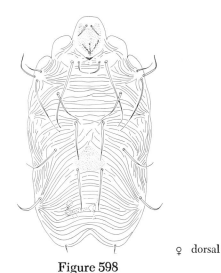

Figure 598

♀ dorsal

♂ dorsal

Figure 599

Figures 598 and 599   *Sciurocoptes tamias* Fain and Hyland

Female with posterior margin incised, propodosomal and opisthosomal shields well developed, opisthosoma small, located in front of setae (d₃), ventral surface of opisthosoma with 22-23 transverse rows of small triangular cuticular scales; epigynium crescentic, well developed, vulva shaped as an inverted "Y"; bursa copulatrix short; all epimera free; male with large hysterosomal shield, shield does not cover posterior part of opisthosoma which bears two lateral, bare and more or less oval areas.

Collected on *Tamias striatus*.

Family Knemidocoptidae

Members of this family are skin parasites of birds and are the cause of a disease commonly known as scaley-legs of chicken. The scaley-leg mite, *Knemidocoptes mutans* (Robin and Lanquetin), can cripple domestic chickens, pheasants, and turkeys. Other members within this family may attack specific regions of the body of a bird. *K. fossor* (Ehlers) is recorded as burrowing into the base of the bill of its host. *Knemidocoptes laevis* var. *gallinae* (Raillet) known as the depluming mite is found attacking the skin of chickens at the base of the feathers on the back, top of wing, around vent on breast and thighs. It results in feather pulling and the loss of feathers over large areas on the body. These mites have been recorded attacking pigeons, pheasants and geese. They are found throughout the United

States. Krantz (1970) placed this species in the genus *Neochemidocoptes* after Fain (1967).

Sclerotized apodemes present beneath scutum; peglike spines absent on idiosoma; tarsi of females with ambulacral suckers .... **Genus *Knemidocoptes***

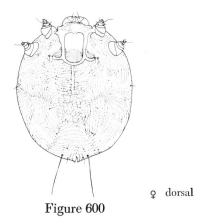

♀ dorsal

Figure 600

Figure 600 *Knemidocoptes mutans* (Robin and Lanquetin)

Anterior transverse genital opening lies parallel with ventral striae, slitlike, genital apodemes absent; anal opening terminal; dorsal striae broken, giving scale or platelike appearance; legs short, stubby, do not end in pretarsi; tarsi clawlike, without long setae; body setae small, a pair of long setae present on posterior margin of body; propodosomal plate present; male with long pretarsi on legs and whiplike setae on tarsi; body setae longer than in female; para-anal suckers absent.

This mite is the cause of scaley leg disease reported in chickens, pheasants, turkeys, partridges, bullfinches, goldfinches, parakeets and other passerine birds. It is probably found throughout the United States due to the shipment of chickens and other birds sold by pet stores.

Keys to the Common Oribatid Beetle Mites of the Cryptostigmata

This section is based entirely on Balogh's 1961, 1963, and 1964 works in which keys are constructed for the world genera of the Oribatid beetle mites. This group of mites is composed of a large number of families and superfamilies that involve approximately 5000 species in some 500 genera. A worker is referred to the work of Krantz (1970) for a list of the superfamilies and families that are contained within the cryptostigmata.

The following keys have been taken from the two works of Balogh (1961, 1964). Only those families and genera recorded from North America or cosmopolitan in distribution have been used. Each of the genera is illustrated by a single species with the species name given. Many of the species have been taken from Balogh (1961, 1964) but are not found in the United States, therefore they are used only as representatives of the genus.

Keys to the Families of the Oribatei
(from Balogh 1961)

1a Tibia and genu more or less of equal length and shape; genital and anal plates meet and fill entire length of ventral side; ventral plate divided into two parts by semicircular or parabolic line, sometimes evanescent in middle; separate adanal and anal plates 2

1b Tibia not shaped like the genu, much shorter; genital and anal plates usually not meeting, not occupying entire length of ventral side; no adanal plate; ventral plate without transversal suture; propodosoma never closable like blade of clasp-knife, never articulated movably 27

2a Legs with two femora; legs not monodactyle in nymphal or larval stages or in part of them; gnathosoma visible from above; notogaster and latero-abdominal gland absent 3

2b Legs with one femur; legs always monodactyle in ontostadia; gnathosoma usually not visible from above; notogaster present 5

3a Sensillus fusiform or claviform .... Family Ctenacaridae

Dorsum with two pairs of long, black hairs; sensillus at least slightly incrassate .... Genus *Ctenacarus* — North America

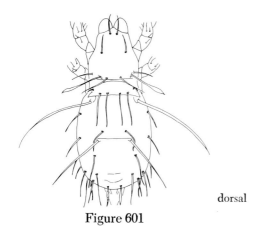

Figure 601

Figure 601 *Ctenacarus araneola* (Grandjean, 1932)

3b Sensillus filiform **4**

4a Asthenic zone large, almost as long as broad; pygidial shield present ....
.... Family Palaeacaridae

Monotypic, with characters of family ....
.... Genus *Palaeacarus* — North America

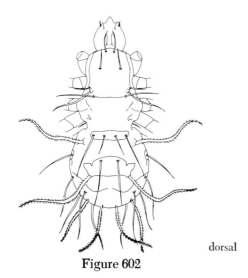

dorsal

Figure 602

Figure 602 *Palaeacarus hystricinus* Tragardh, 1932

4b Asthenic zone considerably shorter than broad; pygidial shield absent ....
.... Family Acaronychidae

Anterior unpaired sclerite bears one pair of hairs only (C_1); hairs C_2 on small sclerite on both sides; hairs e_1 and e_2 black, densely ciliated, much thicker than other hairs ....
.... Genus *Acaronychus* — North America

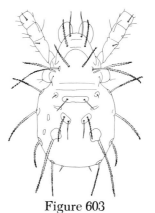

dorsal

Figure 603

Figure 603 *Acaronychus tragardhi* Grandjean, 1932

5a Propodosoma closable like blade of clasp-knife; body usually compressed laterally **6**

5b Propodosoma cannot be closed like blade of clasp-knife; body usually somewhat flattened dorsocentrally **9**

6a Hysterosoma with transversal sutures ....
.... Family Prothoplophoridae

Median claw with tooth below; rostrum dentate; legs tridactylous; lateral claws sometimes very thin, straight, hairlike, claws shorter than tarsus; sensillus fusiform ....
...... Genus *Cryptoplophora* — North America

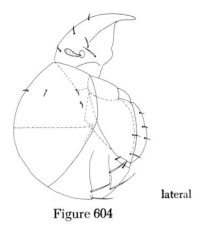

Figure 604

lateral

Figure 604   *Cryptoplophora abscondita*
Grandjean, 1948

6b   Hysterosoma without transversal sutures
.... 7

7a   Ventral plate separate, enclosing genital
and anal plates ....
.... Family Mesoplophoridae

Adanal plates present; anal plates elongated, meeting genital plates along a line ....
.... Genus *Archoplophora* — North America

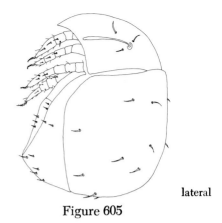

lateral

Figure 605

Figure 605   *Archoplophora rostralis* (Willmann, 1930)

7b   Ventral plate not separate; genital and
anal plates plus eventually aggenital
and adanal plates occupy entire ventral
side 8

8a   Anogenital region wide, not much longer than broad .... Family Phthiracaridae

Interlamellar hair erect; anal plates flat, not protruding even when viewed laterally, with two pairs of anal hairs on inner margins; anogenital region wide, not much longer than broad ....
Genus *Hoplophthiracarus* — North America

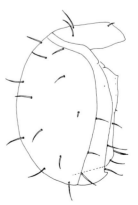

lateral

Figure 606

Figure 606 *Hoplophthiracarus hystricinum* (Berlese, 1908)

8b Anogenital region narrow, considerably longer than broad .... **Family Euphthiracaridae**

Chitinous scale below bothrydium; without aggenital-adanal incisure; aspis without lateral lines .... **Genus *Protoribotritia* — North America**

9a Hysterosoma with from one to three transversal sutures **10**

9b Hysterosoma without transversal sutures **18**

10a Hysterosoma with a single indistinct transversal suture; lateroabdominal gland present .... **Family Parhypochthoniidae**

Hysterosoma with lateral apophysis .... **Genus *Parhypochthonius* — North America**

Figure 607 *Parhypochthonius aphidinus* Berlese, 1904

10b Hysterosoma with one to three distinct transversal sutures; lateroabdominal gland absent **11**

11a Dorsal plate with one suture only (Sometimes there are traces of a second, indistinct suture medially) **12**

11b Dorsal plate with more than one suture **14**

12a Single dorsal plate suture placed between setae c and d; hysterosoma spherical, colorless, with polygonal structure .... **Family Sphaerochthoniidae**

Notogaster circular, with large, polygonal reticulation .... **Genus *Sphaerochthonius* — North America**

dorsal

Figure 608

Figure 608 *Sphaerochthonius transversus* Wallwork, 1960

12b Single dorsal plate suture bearing setae e, or situated closely behind setae row e **13**

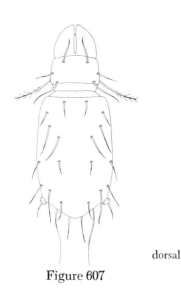

dorsal

Figure 607

13a Setae e or their points of insertion situated on an intercalar sclerite .... Family Hypochthoniidae

Genital plates without oblique transversal suture; notogastral setae not leaf-shaped, shoulder without tubercle .... Genus *Hypochthonius* — North America

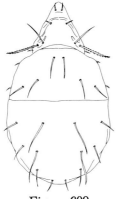

dorsal

Figure 609

Figure 609   *Hypochthonius rufulus* Koch, 1836

13b Setae row e situated on a limbus .... Family Eniochthoniidae

Between suture and prodorsum, an indistinct, medially interrupted, suture-like line .... Genus *Hypochthoniella* — Cosmopolitan

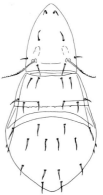

dorsal

Figure 610

Figure 610   *Hypochthoniella minutissima* (Berlese, 1904)

14a Sutures without intercalar sclerites 15

14b One or more sutures with intercalar sclerites bearing setae 16

15a Two sutures present between setal rows c and d and between e and f .... Family Brachychthoniidae

Suprapleural plates four in number, second pleural plate free; setae d_3 marginally situated, separated from tergites by longitudinal suture .... **Genus *Eobrachychthonius* — North America**

lateral

Figure 611

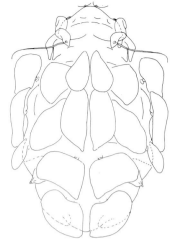

dorsal

Figure 613

Figure 613 *Pterochthonius angelus* (Berlese, 1910)

16a "Eyes" present, three in number; some setae on dorsum very long and smooth .... Family **Heterochthoniidae**

Prodorsum with three well discernible "eyes," notogastral hairs smooth ....
.... **Genus** *Heterochthonius* — **North America**

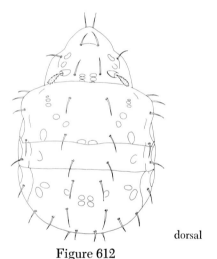

dorsal

Figure 612

Figures 611 and 612 *Eobrachychthonius latior* (Berlese, 1910)

15b Three sutures present; between c and d, d and e, and e and f; dorsal setae leaf-shaped .... Family **Pterochthoniidae**

All notogastral hairs dilatate, three sutures closely adjacent ....
...... **Genus** *Pterochthonius* — **North America**

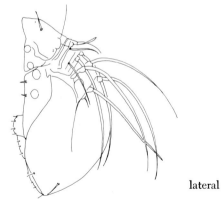

lateral

Figure 614

Figure 614   *Heterochthonius gibbus* Berlese, 1910

16b   "Eyes" absent; dorsal setae ciliated or leaf-shaped 17

17a   Setae f_1 and f_2 erectile, born on intercalar sclerite; tergite D not sclerotized; setae d absent .... Family Atopochthoniidae

Some notogastral setae blade-shaped, considerably longer, than wide .... Genus *Atopochthonius* — North America

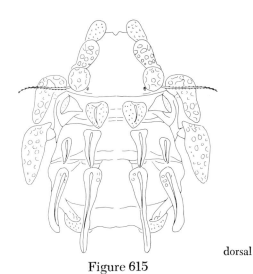

dorsal

Figure 615

Figure 615   *Atopochthonius artiodactylus* Grandjean, 1948

17b   Setae e and f erectile, borne on intercalar sclerites, tergite D sclerotized .... Family Cosmochthoniidae

Legs I bidactylous; legs II-IV tridactylous .... Genus *Cosmochthonius* — Cosmopolitan

dorsal

Figure 616

Figure 616   *Cosmochthonius reticulatus* Grandjean, 1947

18a   Propodosoma and hysterosoma movably connected 19

18b   Propodosoma and hysterosoma immovably connected 22

19a   Parabolic transversal suture behind genital plate (pseudodiagastry); large, pleotrichial epimeres .... Family Eulohmanniidae

A parabolic suture between genital and anal plates .... Genus *Eulohmannia* — North America

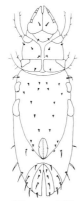

ventral

Figure 617

Figure 617   *Eulohmannia ribagai* Berlese, 1910

19b   No parabolic transversal suture behind genital plate **20**

20a   A horizontal suture behind genital plate (schizogastry); latero-abdominal gland present .... Family **Epilohmanniidae**

Horizontal suture behind genital plate; latero-abdominal gland present ....
.... Genus **Epilohmannia** — **Cosmopolitan**

dorsal

Figure 618

Figure 618   *Epilohmannia cylindrica* (Berlese, 1905)

20b   Horizontal suture absent behind genital plate **21**

21a   Body more or less cylindrical; margin of hysterosoma strongly convex, posteriorly .... Family **Lohmanniidae**

Pygidial neotrichy absent; sensillus filiform, pectinated; fossulae present; anal and adanal plates separated ....
.... Genus **Lohmannia** — **Cosmopolitan**

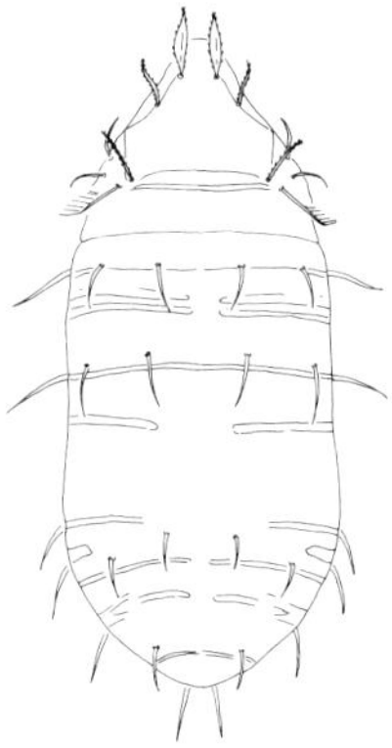

Figure 619

dorsal

Figure 619   *Lohmannia javena* Balogh, 1961

21b   Body more or less flattened; posterior border of propodosoma narrower than anterior margin of hysterosoma ....
.... Family **Perlohmanniidae**

Legs monodactyle ....
.... Genus **Perlohmannia** — **North America**

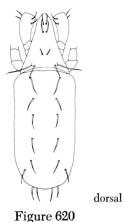

Figure 620

Figure 620 *Perlohmannia dissimilis* (Hewitt, 1908)

22a A medially interrupted semicircular suture behind genital plate (diagastry); hysterosoma cylindrical .... **Family Nanhermanniidae**

Notogastral hairs bifid, branches flagelliform; body dirty, bearing minute particles of earth; genital and anal plates not meeting each other; a semicircular medially interrupted suture between genital and anal plates on ventral side (diagastry) .... **Genus Masthermannia — Cosmopolitan**

Figure 621

Figure 621 *Masthermannia mamillaris* (Berlese, 1904)

22b Ventral plate without transversal suture (not diagastry) **23**

23a Genital and anal plates on separate ventral plate; body round; shape and structure of genu and tibia rather diverse .... **Family Hermanniidae**

Notogastral hairs not distending like knifeblade .... Genus Hermannia — North America

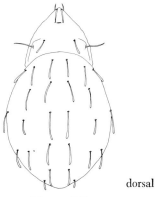

Figure 622

Figure 622 *Hermannia gibba* (C. L. Koch, 1839)

23b Genital and anal plates occupy entire ventral side, no separate ventral plate **24**

24a Bothrydium absent; a special, porose, birefringent cerotegument .... **Family Malaconothridae**

Legs tridactylous; latero-abdominal gland absent; cerotegument porose, birefringent .... Genus Trimalaconothrus — Cosmopolitan

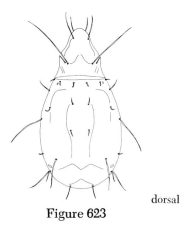

dorsal

Figure 623

Figure 623 *Trimalaconothrus glaber* (Michael, 1888)

24b Bothrydium present; if absent, then no birefringent cerotegument present 25

25a Epimeres with neotrichy; aggenital setae absent, only a part of genital setae in a marginal position; genital plates without neotrichy ...... Family Nothridae

Epimeres with neotrichy; only some genital setae in marginal position .... Genus *Nothrus* — Cosmopolitan

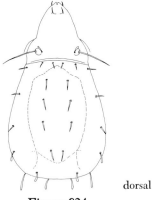

dorsal

Figure 624

Figure 624 *Northrus partensis* Sellnick, 1928

25b Epimeres without neotrichy; genital setae conspicuously in marginal position; occasionally neotrichy on genital plates 26

26a Two pairs of aggenital setae; dorsum flat or concave .... Family Camisiidae

Hysterosoma with more or less parallel sides, broadly truncate posteriorly, with or without apophyses .... Genus *Camisia* — North America

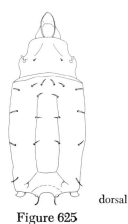

dorsal

Figure 625

Figure 625 *Camisia horrida* (Hermann, 1804)

26b Aggenital setae absent; dorsum convex .... Family Trhypochthoniidae

Legs monodactyle; 18-20 pairs of genital setae; rostral setae originating near each other .... Genus *Mucronothrus* — North America

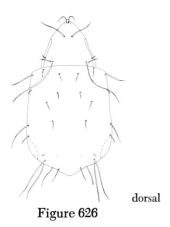

dorsal

Figure 626

Figure 626   *Mucronothrus rostratus* Tragardh, 1831

27a Hysterosoma with pteromorphae bending downward or horizontal, if horizontal then areae porosae, sacculi, or pori always present on dorsum 28

27b Hysterosoma without pteromorphae, if species have a horizontal appendage (genera *Xenillus, Hafenferrefia*), resembling horizontal pteromorphae, it extends anteriorly, never laterally, dorsum in such cases always pycnonotic 38

28a Pteromorphae earlike, extending far anteriorly and posteriorly 29

28b Pteromorphae never earlike, if superficially earlike, then never extending far anteriorly and posteriorly 30

29a Six pairs of genital setae; lamellae reduced to a line or absent; area porose usually present; pteromorphae mostly with fissure; prodorsum without large, broad lamellae ...... Family Galumnidae

Legs tridactylous; one median pore on notogaster .... Genus *Allogalumna*—Cosmopolitan

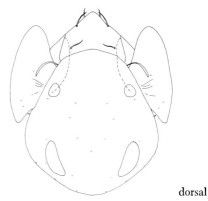

dorsal

Figure 627

Figure 627   *Allogalumna confluens* Balogh, 1960

29b Four or five pairs of genital setae; lamellae situated marginally, always present; pteromorphae without fissure .... Family Parakalummidae

Ventral margin of pteromorphae slightly concave anteriorly; lamellae linear; anterior margin of pteromorphae rounded .... Genus *Neoribates* — North America

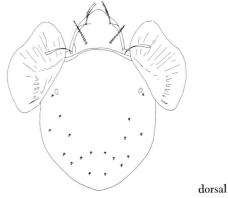

dorsal

Figure 628

Figure 628   *Neoribates aurantiacus* (Oudemans, 1913)

30a   Interlamellar setae leaflike; body with wide and thick cerotegument .... Family Pelopidae

Interlamellar setae short, setaelike, anterior margin of pteromorphae more projecting than dorsosejugal suture ....Genus *Peloptulus* — North America

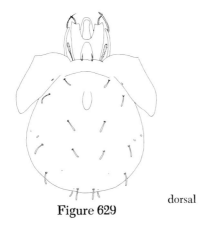

Figure 629   dorsal

Figure 629   *Peloptulus foveolatus* Hammer, 1961

30b   Interlamellar setae never leaf-shaped; usually no thick cerotegument and fusiform setae 31

31a   Lamellae either narrow slits along margin of prodorsum or if broader, never meeting in middle, sometimes connected by a translamella leaving a part of prodorsum free 33

31b   Lamellae very broad, usually meeting or fusing in middle, largely covering prodorsum 32

32a   Lamellae entirely fused in middle, forming a single large scale almost entirely covering prodorsum .... Family Tegoribatidae

True areae porosae present; anterior tip of pteromorphae much farther projecting than middle or dorsosejugal suture .... Genus *Lepidozetes* — North America

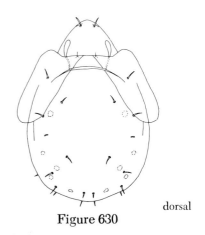

Figure 630   dorsal

Figure 630   *Lepidozetes singularis* Berlese, 1910

32b   Lamellae only meet in median line or fuse basally, with apices remaining free .... Families Achipteriidae, Oribatellidae

Family Achipteriidae
Notogaster with true areae porosae .... Genus *Parachipteria* — North America

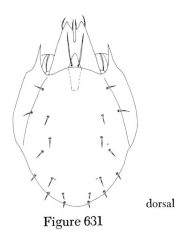

dorsal

Figure 631

Figure 631 *Parachipteria punctata* (Nicolet, 1855)

Family Oribatellidae
Interlamellar setae very short; external point of lamellae much shorter than width of lamellae ....
...... Genus *Ophidiotrichus* — North America

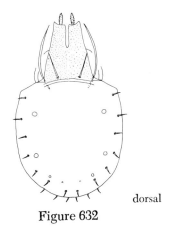

dorsal

Figure 632

Figure 632 *Ophidiotrichus connexus vindobonensis* Piffl, 1961

33a One of the following features will be present: a short, horizontal pteromorpha present; lamellae tapering anteriorly,

without cuspus; one to five pairs of genital setae; dorsosejugal suture of three arches ....
Families Oribatulidae, Haplozetidae, Oripodidae

Family Oribatulidae
Lamellae ribbon-shaped; long cuspis frequently widening; legs tridactylous ....
.... Genus *Zygoribatula* — Cosmopolitan

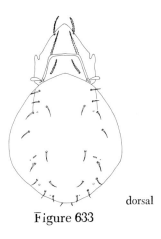

dorsal

Figure 633

Figure 633 *Zygoribatula frisiae* (Oudemans, 1900)

Family Haplozetidae
Pteromorphae immovable; legs tridactylous; one pair of aggenital setae ....
.... Genus *Peloribates* — Cosmopolitan

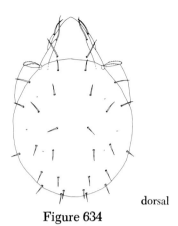

dorsal

Figure 634

Figure 634　*Peloribates europaeus* Willmann, 1935

Family Oripodidae
One to three pairs of genital setae; two pairs of adanal setae ....
...............Genus *Exoribatula* — North America

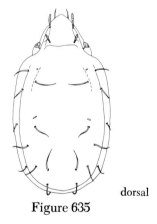

dorsal

Figure 635

Figure 635　*Exoribatula flagelligera* (Balogh, 1958)

33b　**True pteromorphae bending downward; lamellae with parallel sides; six or more genital hairs** 34

34a　**Hysterosoma very broad, broader than long, A_2 and A_3 either very long and ribbonlike, or divided into several round portions** .... Family Mochlozetidae

Notogaster without setae, without chitinous bridge between pteromorphae, circular, as long as broad; translamella ribbonlike; small species .... Genus *Podoribates*—Cosmopolitan

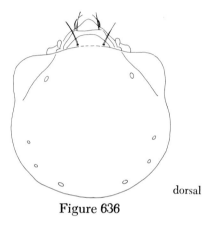

dorsal

Figure 636

Figure 636　*Podoribates longipes* (Berlese, 1887)

34b　**Hysterosoma not broader than long, A_2 and A_3 not long and ribbonlike, not divided into several portions** 35

35a　**Lamellar setae originate anteriorly to cuspis on surface of prodorsum; cuspis pointed; dorsal setae absent** ....
.... Family Chamobatidae

Lamellae not attenuated, with cuspis; interlamellar setae not conspicuously long ....
.... Genus *Chamobates* — Cosmopolitan

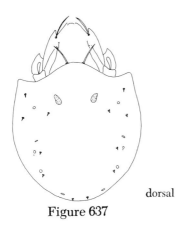

dorsal

Figure 637

Figure 637 *Chamobates* sp.

35b Lamellar seta originate on cuspis of lamella **36**

36a Pteromorphae narrow (lateral view), extending deep downward, pyconotic; found in wet localities, in *Sphagnum* .... Family **Limnozetidae**

Lamellae straight, with short cuspides .... Genus *Limnozetes* — Cosmopolitan

dorsal

Figure 638

Figure 638 *Limnozetes canadensis* Hammer

36b Pteromorphae triangular (lateral view), short, mostly poronotic **37**

37a Pteromorphae movable, connected by chitinous bridge, which covers base of prodorsum like eaves .... Family **Mycobatidae**

Translamella present or absent; dorsosejugal suture and interlamellar setae present; cuspis much shorter than lamella; 10 or 11 pairs of notogastral setae; anterior margin of pteromorphae arching farther forward than middle of dorsosejugal suture; pteromorphae connected by narrow chitinous bridge .... Genus *Mycobates* — North America

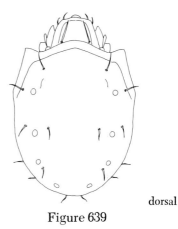

dorsal

Figure 639

Figure 639 *Mycobates tridactylus* Willmann, 1929

37b Pteromorphae not movable; chitinous bridge absent ...... Family **Ceratozetidae**

Translamella present or absent; pseudotectum one without teeth; sensillus short; notogastral setae very small or absent, or long, between cuspides; rostrum without round process in middle .... Genus *Ceratozetes* — North America

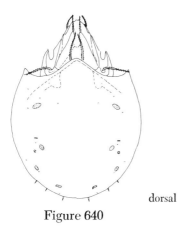

dorsal

Figure 640

Figure 640   *Ceratozetes peritus* Grandjean, 1951

dorsal

Figure 641

Figure 641   *Passalozetes africanus* Grandjean, 1932

38a Poronotic mites; two to six pairs of areae porosae; sacculi, or pori in dorsum, sometimes minute, hard to find on sculptured or dark species 39

38b Pycnonotic mites 42

39a No lamellae or translamella, at most, basal portion of lamellae present; lenticulus circular; two pairs of pointlike areae porosae .... Family Passalozetidae

Notogaster with circular lenticulus; bothrydium and sensillus well developed .... **Genus** *Passalozetes* — Cosmopolitan

39b Lamellae and, at times, translamella present 40

40a Sensillus fan-shaped; dorsosejugal suture pointed anteriorly; two pairs of dotlike areae porosae .... **Family Licneremaeidae**

Sensillus flabelliform or licheniform; porodorsum with lamellae or costulae; notogaster with rough sculpture, without large glandule on each side; bothrydium present .... **Genus** *Licneremaeus* — North America

dorsal

Figure 642

Figure 642 *Licneremaeus prodigiosus* Schuster, 1958

40b Sensillus not fan-shaped; dorsosejugal suture not pointed; at least three pairs of areae porosae, sacculi, or pori **41**

41a Lenticulus present, quadrangular; lamellae and translamella present .... Family **Scutoverticidae**

Lamellae with cuspis; notogaster with quadrangular lenticulus .... Genus **Scutovertex** — Cosmopolitan

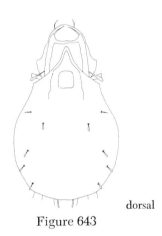

dorsal

Figure 643

Figure 643 *Scutovertex minutus* (C. L. Koch, 1836)

41b Lenticulus absent; hysterosoma normally not circular, mostly not foveolated; one to five pairs of genital setae, if six pairs of genital setae, then true areae porosae present .... Superfamily .... **Oribatuloidea** (see couplet 33)

42a True lamellae present, these may be flat, lathlike, bearing lamellar setae .. **43**

42b True lamellae absent, at most narrow costulae present **55**

43a Shoulder with a pointed, pteromorphalike appendage extending forward; dorsum not sculptured .... Family **Tenuialidae**

Apex of humeral appendage dentate; lamellae wide and long; shoulder with proclinate triangular appendage .... Genus **Tenuiala** — North America

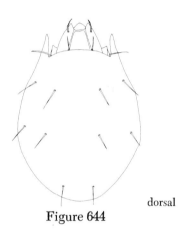

dorsal

Figure 644

Figure 644 *Tenuiala nuda* Ewing, 1913

43b Shoulder without pteromorphalike appendage, if such an appendage appears to be present, then dorsum always sculptured **44**

44a Lamellae either situated parallel to each other in middle of prodorsum, or converge until apices meet **45**

44b Lamellae never beside each other, their apices never meeting, at most connected by translamella **49**

45a Lamellae situated more or less parallel to each other in middle of prodorsum; body covered by cerotegument and adhering particles of dirt; end of lamellae attenuated, pointed; dorsum sculptured, oval .... Part of Cepheidae

Body covered with cerotegument and adhering particles of dirt; cuspides foot-shaped, extending beyond rostrum ....
.... Genus *Eupterotegaeus* — North America

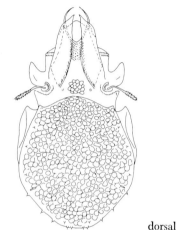

Figure 645   dorsal

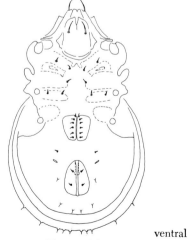

Figure 646   ventral

Figures 645 and 646   *Eupterotegaeus rhamphosus* Higgins and Woolley, 1968

45b Lamellae covered anteriorly in an "A" shape, apices meeting 46

46a Dorsum smooth, at most finely granulated or punctate 47

46b Dorsum sculptured (foveolate, reticulate, or rugose) 48

47a Genital and anal plates very large, almost touching each other ....
.... Family Astegistidae

Proclinate triangular appendage on shoulder absent; lamellar cuspides meeting; legs monodactyle .... Genus *Cultroribula—Cosmopolitan*

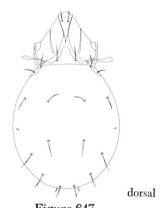

dorsal
Figure 647

Figure 647   *Cultroribula argentinensis* Balogh and Csiszar, 1963

47b Genital and anal plates removed from each other; with two pairs of simple humeral setae; ten additional pairs of notogastral setae, six genital setae; legs II, III, IV inserted somewhat medially and remote from lateral margins of body .... **Family Liacaridae**

Sensillus spindleform .... **Genus *Liacarus* — North America**

Sensillus clavate, barbed .... **Genus *Xenillus* — North America**

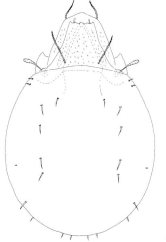

dorsal

Figure 650

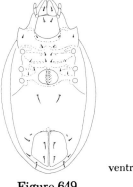

dorsal

Figure 648

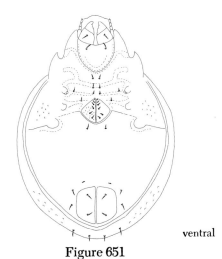

ventral

Figure 651

ventral

Figure 649

Figures 648 and 649 *Liacarus cidarus* Woolley, 1968

Figures 650 and 651 *Xenillus gelasillus* Woolley and Higgins, 1966

48a Integument pitted or rugose; lamellae with claviform or spindleform sensilli; two humeral notogastral bristles, five or six pairs of genital setae; tuberculous trochanteral fossae II, III .... **Family Xenillidae**

48b Not as above; six pairs of genital setae; adjacent minute setae absent on shoulder; dorsum reticulate or rugose .... **Family Cepheidae (Part of)**

Tarsus tridactylous; notogaster with roughly sculptured shoulders ....
.... Genus *Sphodrocepheus* — North America

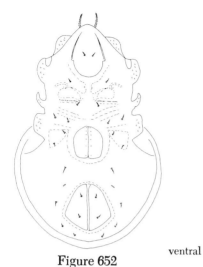

Figure 652   ventral

Figure 652   *Sphodrocepheus anthelionus* Woolley and Higgins, 1968

49a Lenticulus present, oval or quadrangular, with rounded corners ....
.... Family Scutoverticidae

Notogaster with lenticulus; interlamellar setae short or absent ....
.... Genus *Scutovertex* — Cosmopolitan

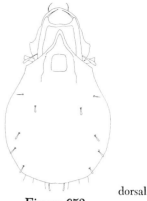

dorsal

Figure 653

Figure 653   *Scutovertex minutus* (C. L. Koch, 1836)

49b Lenticulus absent 50

50a Dorsum sculptured; hysterosoma circular or oval, reticulate or rugose ....
.... Family Cepheidae (Part of)

Sensillus spherical, stalkless, situated in bothrydium; six pairs of genital hairs; without translamella ....
......Genus *Ommatocepheus* — North America

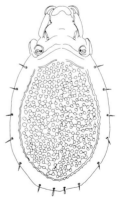

dorsal

Figure 654

Figure 654 *Ommatocepheus pulcherrimus* Berlese, 1913

50b Dorsum smooth or at most punctate or finely granulated (some species covered with coarsely granulated or rugose cerotegument 51

51a Mandible long, bacilliform, without chelae, apex serrate; hysterosoma round .... Family Gustaviidae

Mandibles very long, bacilliform, without chelae, apex serrate ...., ... Genus *Gustavia* — North America

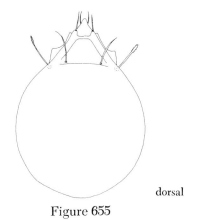

dorsal

Figure 655

Figure 655 *Gustavia microcephala* (Nicolete, 1855)

51b Mandible always chelated, sometimes pleoptoid 52

52a Legs IV thicker than other legs, with thick spine .... Family Zetorchestidae

No chitinous ridges extending backward from base of rostral setae; legs IV thicker than other legs; hysterosoma almost circular in dorsal view .... Genus *Zetorchestes* — North America

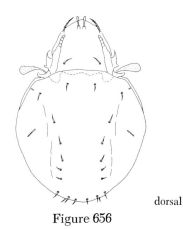

dorsal

Figure 656

Figure 656 *Zetorchestes flabrarius* Grandjean, 1951

52b Leg IV not thickened 53

53a Notogaster with protruding humeral apophysis; dorsum covered sometimes with granulated cerotegument; interlamellar setae minute or absent .... Family Tectocepheidae

Lamellae separated medially, without a complete translamella; sensillus clavate, spined; six pairs of genital setae; legs with large spines .... Genus *Exochocepheus* — North America

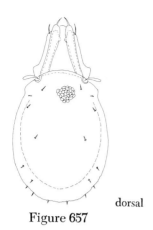

dorsal

Figure 657

Figure 657   *Exochocepheus eremitus* Woolley and Higgins, 1968

53b Notogaster without protruding humeral apophysis; dorsum not covered with granulated cerotegument 54

54a Translamella present, linear, 800-900 microns in size .... Family Cepheidae (part of)

Notogaster smooth, notogastral setae very small or absent; with translamella, lamellae not parallel, never meeting apically .... Genus *Conoppia* — North America

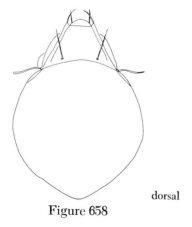

dorsal

Figure 658

Figure 658   *Conoppia microptera* (Berlese, 1885)

54b Translamella absent (if present, species smaller than 500 microns) .... Family Metrioppiidae

Mandibles peloptoid .... Genus *Metrioppia* — North America

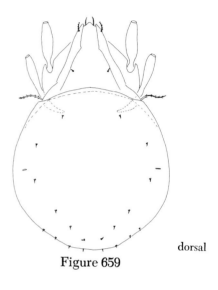

dorsal

Figure 659

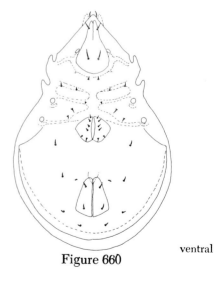

ventral

Figure 660

Figures 659 and 660 *Metrioppia oregenensis* Woolley and Higgins, 1969

55a Genital plates with transversal sutures, usually bearing excentrically situated exuviae .... Family Liodidae

Notogaster flat; ventral plate not closed behind anal plate ....
...............Genus *Platyliodes* — North America

dorsal

Figure 661

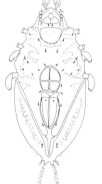

ventral

Figure 662

Figures 661 and 662 *Platyliodes macropionus* Woolley and Higgins, 1969

55b Genital plates without transversal suture 56

56a Lateral tubus on sides of body ....
.... Family Hermanniellidae

Bothrydia originating far from each other, near margin of prodorsum; prodorsum with lateral carinae imitating lamellae ....
.... Genus *Hermanniella* — Cosmopolitan

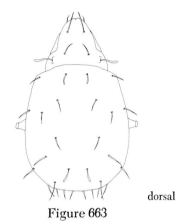

dorsal

Figure 663

Figure 663 *Hermanniella granulata* (Nicolet, 1855)

56b No lateral tubus on sides of body 57

57a Ventral plate with more than four pairs of setae 58

57b Ventral plate with only four pairs of setae 59

58a Anal plate with three to five pairs of setae; sensillus never distending like a fan or a lichen, if sensillus slightly distended, then legs conspicuously long and tridactylous .... Family Eremaeidae

10 or 11 pairs short, or medium notogastral setae ...... Genus *Eremaeus* — North America

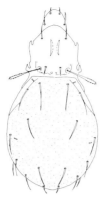

dorsal

Figure 664

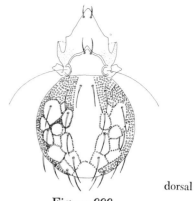

dorsal

Figure 666

Figure 666   *Eremobelba geographica* Berlese, 1908

ventral

Figure 665

Figures 664 and 665   *Eremaeus silvestris* Forsslund, 1956

58b   Anal plate with two pairs of setae; dorsum without exuviae; dorsosejugal suture sharp, conspicuous; dorsum with granulated cerotegument; prodorsum always with costula; epimeres often with star-shaped setae ....
.... **Family Eremobelbidae**

Costulae short; sensillus not pectinate ....
.... **Genus *Eremobelba* — Cosmopolitan**

59a   Legs very long, if not much longer than body, like a string of pearls or bearing articulatory capsules, always tridactylous; hysterosoma either spherical or circular **60**

59b   Legs not conspicuously long; hysterosoma never spherical but oval; not bearing two longitudinal rows of 8 pairs of dorsal setae **61**

60a   Legs tridactylous, long, filiform, if seemingly short, then with articulatory capsules; hysterosoma circular, dorsum often flat, carrying exuviae; at most with six pairs of posteromarginal setae ....
.... **Family Plateremaeidae**

Four to six pairs of anal setae ....
.... **Genus *Plateremaeus* — North America**

dorsal

Figure 667

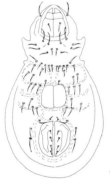

ventral

Figure 668

Figures 667 and 668 *Plateremaeus mirabilis* Csiszar, 1962

60b Legs monodactyle, like string of pearls, often very long, always without articulatory capsules; hysterosoma mostly spherical, with two longitudinal rows of eight pairs of dorsal and three pairs of posteromarginal setae .... Family Damaeidae

Lateral appendage between leg I and II of propodosoma unicuspidate; spinae adnatae present; dorsosejugal suture arched ....Genus *Epidamaeus* — North America

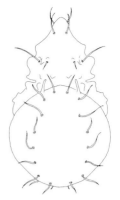

dorsal

Figure 669

Figure 669 *Epidamaeus flexispinosus* Kunst, 1961

61a 13-15 pairs of dorsal setae present **62**

61b No more than 11 pairs of dorsal setae present **65**

62a Dorsal plate distends bilaterally to ventral surface; ventral plate V-shaped; dorsum coarsely reticulated, sometimes broadly marginate **63**

62b Dorsal plate not distending bilaterally to ventral surface; ventral plate not V-shaped **64**

63a 13 pairs of dorsal setae .... Family Cymbaeremaeidae

Marginal zone of notogaster distinct from central area; six pairs of genital setae; notogaster rugose .... Genus *Scapheremaeus* — Cosmopolitan

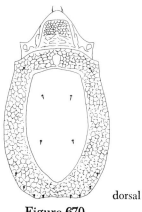

dorsal

Figure 670

Figure 670 *Scapheremaeus palustris* Sell-nick, 1928

63b 14 pairs of dorsal setae ....
.... Family Micreremidae

Four pairs of genital setae; 14 pairs of noto-gastral setae; notograster with polygonal structure; dorsosejugal suture not interrupted medially ....
.... Genus *Micreremus* — North America

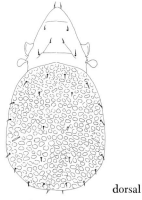

dorsal

Figure 671

Figure 671 *Micreremus brevipes* (Michael, 1888)

64a Dorsum smooth or finely punctate, never granulated or rugose; legs monodactyle; fresh water species; 13 pairs of dorsal setae .... Family Hydrozetidae

Without pteromorphae and translamella ....
.... Genus *Hydrozetes* — Cosmopolitan

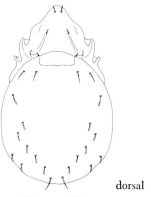

dorsal

Figure 672

Figure 672 *Hydrozetes confervae* (Schrank, 1780)

64b Dorsum granulated, rugose or covered by cerotegument; legs tridactylous or monodactylous ....
Families Ameronothridae and Podacaridae

Family Ameronothridae
Lamellae present; two exobothridial setae; absent of a dorsosegugal suture ....
.... Genus *Caenosamerus* — North America

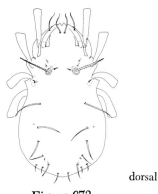

dorsal

Figure 673

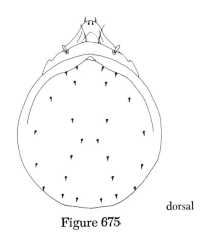

dorsal

Figure 675

Figure 675 *Alaskozetes coriaceus* Hammer, 1955

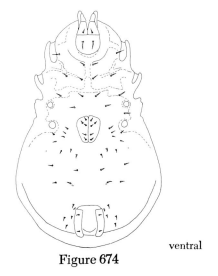

ventral

Figure 674

Figures 673 and 674 *Caenosamerus litosus* Higgins and Woolley, 1969

Family Podacaridae
Dorsosejugal suture evenly curved; sensillus short, dilatate; interlamellar suture long .... Genus *Alaskozetes* — North America

65a Mandibles peloptoid; chelicerae without teeth; prodorsum often with oval hollow and sharply incumbent rostral setae .... Family Suctobelbidae

Rostrum without lateral rostral teeth, elongated into a snout or nose with two large lateral horns, sensillus elongated; clavate; anterior margin of notogaster entire without toothlike projections; six pairs of genital setae .... Genus *Rhinosuctobelba* — North America

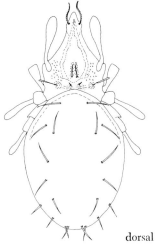

dorsal

Figure 676

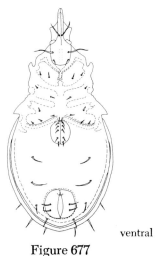

ventral

Figure 677

Figures 676 and 677 *Rhinosuctobelba dicerosa* Woolley and Higgins, 1969

65b Mandibles not peloptoid, normal; rostrum not strikingly narrow and pointed 66

66a Elongated species, with three or four genital and nine to ten pairs of dorsal setae; with chitinous tubercles or ridges on border of prodorsum and hysterosoma; body not flattened, mostly finely punctate; prodorsum with longitudinal costula on both sides; genital plate darker than ventral plate ....
.... Family Otocepheidae

Pedotecta two not strikingly large, four separated tubercles on anterior margin of notogaster ....
.... Genus *Dolicheremaeus* — Cosmopolitan

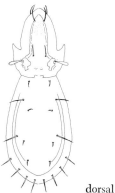

dorsal

Figure 678

Figure 678 *Dolicheremaeus clavatus* (Aoki, 1959)

66b Not strikingly elongated species, elongated in appearance then no conspicuous chitinous tubercles or ridges on border of prodorsum and hysterosoma 67

67a Leg IV with thick spines used for jumping; a chitinous ridge on hysterosoma, directed backwards from shoulder; rostral setae originating close to each other, plumose or flabelliform, or apically bifurcated ....
.... Family Zetorchestidae (See couplet 52)

67b Leg IV never utilized for jumping; no ridge on hysterosoma; rostral setae not originating close to each other 68

68a Genital and anal plates very large, almost touching; costulae usually double .... Family Thyrisomidae

Coxisternal ridge three present; genital and anal plates large; chelicerae not peloptoid .... Genus *Oribella* — North America

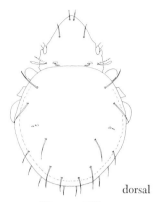

dorsal

Figure 680

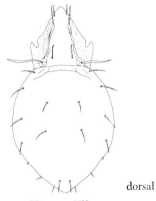

dorsal

Figure 679

Figure 679   *Oribella cavatica* Kunst, 1962

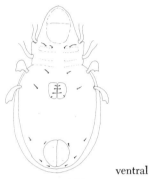

ventral

Figure 681

Figures 680 and 681   *Oppia coloradensis* Woolley 1969

68b Genital and anal plates not strikingly large; costulae never double; 9-11 pairs of dorsal setae in medial and marginal rows; coarse reticulation absent .... Families Oppiidae and Autognetidae

Family Oppiidae
Anterior margin of notogaster with one to three pairs of longitudinal chitinous carinae or lines; prodorsum without long, parallel costulae .... Genus *Oppia* — Cosmopolitan (part of)

Family Autognetidae
Sensillus setiform, denticulate on one side; notogastral setae of nymphs broad, leaf- or fan-shaped .... Genus *Conchogneta* — North America

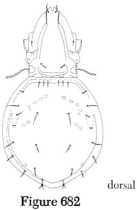

dorsal

Figure 682

Figure 682 *Conchogneta dalecarlica* (Forsslund, 1947)

Selected References

Arthur, D. R. 1962. Ticks and Disease. Pergamon Press, Oxford: 445 pp.

Atyeo, W. T. 1960. A revision of the mite family Bdellidae in North and Central America. Univ. Kansas Sci. Bull. 40(8):345-499.

Atyeo, W. T., E. W. Baker and D. A. Crossley, Jr. 1961. The genus *Raphignathus* Duges (Acarina, Raphignathidae) in the United States with notes on the Old World species. Acarologia 3(1):14-20.

Atyeo, W. T. and E. W. Baker. 1964. Tarsocheylidae, a new family of prostigmatic mites (Acarina). Bull Univ. Nebraska State Mus. 4(11):243-256.

Atyeo, W. T. and N. L. Braasch. 1966. The feather mite genus *Proctophyllodes* (Sarcoptiformes: Proctophyllodidae). Bull. Univ. Nebraska State Mus. 5:1-354.

Atyeo, W. T. and J. Gaud. 1966. The chaetotaxy of sarcoptiform feather mites (Acarina: Analgoidea). Jour. Kansas Ent. Soc. 39 (2):337-346.

Baker, E. W. and A. Hoffmann. 1948. Acaros de la familia Cunaxidae. Anales Esc. Nac. Ciencias Biol. 5(3-4):229-273.

Baker, E. W. 1949. A review of the mites of the family Cheyletidae in the United States National Museum. Proc. U.S. Nat. Mus. 99(3238):267-320.

Baker, E. W. and G. W. Wharton. 1952. An introduction to Acarology. The Macmillan Co., New York: 465 pp.

Baker, E. W., T. M. Evans, D. J. Gould, W. B. Hull and H. L. Keegan. 1956. A manual of parasitic mites of medical or economic importance. Nat. Pest Cont. Assoc. Tech. Publ.: 170 pp.

Baker, E. W., J. H. Camin, F. Cunliffe, T. A. Wooley and C. E. Yunker. 1958. Guide to the families of mites. Inst. Acarology Contrib. 3:1-242.

Baker, E. W. 1965. A review of the genera of the family Tydeidae (Acarina). Advances in Acarology 2:95-133. Cornell Univ. Press. Ithaca, N.Y.

Banks, N. 1915. The Acarina or mites. U.S. Dept. Agr. Rept. 108:1-153.

Brennan, J. M. and E. K. Jones. 1959. Keys to the chiggers of North America with synonymic notes and descriptions of two new genera (Acarina: Trombiculidae). Ann. Ent. Soc. Amer. 52(1):7-16.

Camin, J. H. and F. Gorirossi. 1955. A revision of the suborder Mesostigmata (Acarina), based on new interpretations of comparative morphological data. Chicago Acad. Sci. Spec. Publ. 11:70 pp.

Chant, D. A. 1965. Generic concepts in the family Phytoseiidae (Acarina: Mesostigmata). Can. Ent. 97(4):351-374.

Clark, G. M. 1964. The acarine genus *Syringophilus* in North American birds. Acarologia 6(1):77-92.

Cooreman, J. 1954. Acariens Canestriniidae de la collection A. C. Oudemans, a Leiden. Zool. Meded. 33(13):83-90.

Cross, E. A. 1965. The generic relationships of the family Pyemotidae (Acarina: Trombidiformes). Univ. Kansas Sci. Bull. 45 (2):29-275.

Crossley, D. A. 1960. Comparative external morphology and taxonomy of nymphs of the Trombiculidae (Acarina). University of Kansas Sci. Bull. 40(6):135-321.

Dubinin, W. B. 1951. Feather mites (Analgesoidea). Part I. Introduction to their study. Fauna U.S.S.R. 6(5):1-363.

Dubinin, W. B. 1953. Feather mites (Analgesoidea). Part II. Epidermoptidae and Freyanidae. Fauna U.S.S.R. 6(6):1-411.

Dubinin, W. B. 1956. Feather mites (Analgesoidea). Part III. Pterolichidae. Fauna U.S.S.R. 6(7):1-814.

Evans, G. O. 1957. An introduction to the British Mesostigmata (Acarina) with keys to the families and genera. Linn. Soc. Jour. Zool. 43:203-259.

Evans, G. O., J. G. Sheals and D. Macfarland. 1961. The terrestrial Acari of the British Isles. Vol. I. Introduction and biology. British Museum, London: 219 pp.

Evans, G. O. and W. M. Till. 1966. Studies on the British Dermanyssidae (Acari: Mesostigmata). Part Classification. Bull. Brit. Mus. (Nat. Hist.) Zool. 14(5):109-370.

Fain, A. and J. Bafort. 1964. Les Acariens de la famille Cytoditidae (Sarcoptiformes). Description de sept especes nouvelles. Acarologia 6(3):504-528.

Gaud, J. and J. Mouchet (1957). Acariens plumicoles (Analgesoidea) des oiseaux du Cameroun. I. Proctophyllodidae. Ann. Parasit. Hum. Comp. 32(5-6):491-546.

Higgins. H. G. and S. B. Mulaik. 1957. Another *Caeculus* from southwestern United States (Caeculidae). Texas J. Sci. 9:267-269.

Higgins, H. G. and S. B. Mulaik. 1961. Additional distribution records of North American rake-legged mites. Proc. Ent. Soc. Wash. 63:209-210. (Caeculidae).

Hirst, S. 1919. The genus *Demodex* Owen. Brit. Mus. (Nat. Hist.) Studies on Acari 1:1-44.

Hughes, A. M. 1961. The mites of stored food. Minis. Agr., Fish and Food Tech. Bull. 9:287 pp. + vi.

Hughes, R. D. and C. G. Jackson. 1958. A review of the Anoetidae (Acari). Virginia Jour. Sci. 9, N.S. (1):5-198.

Hughes, T. E. 1959. Mites or the Acari. Univ. of London, Athlone Press: 225 pp.

Jameson, E. W. 1955. A summary of the genera of Myobiidae (Acarina). Jour. Parasit. 41(4):407-416.

Jeppson, L. R., H. H. Keifer and E. W. Baker. 1975. Mites Injurious To Economic Plants, Univ. of Californa Press: 614 pp.

Johnston, D. E. 1964. *Psorergates bos*, a new mite parasite of domestic cattle (Acari-Psorergatidae). Ohio Agr. Exp. Sta. Res. Circ. 129:7 pp.

Johnston, D. E. 1968. An atlas of Acari I. The families of Parasitiformes and Opilioacariformes. Acar. Lab., Ohio State Univ. 172:110 pp. + x.

Keifer, H. H. 1952. The eriophyid mites of California (Acarina: Eriophyidae). Bull. Calif. Insect Surv. 2(1):123 pp.

Krantz, G. W. 1970. A manual of Acarology. O.S.U. Book Stores Inc., Corvallis, Oregon: 335 pp.

Krantz, G. W. 1978. A manual of Acarology, sec. ed. O.S.U. Book Stores Inc. Corvallis, Oregon: 509 pp.

Krombein, K. V. 1961. Some symbiotic relations between saproglyphid mites and solitary vespid wasps (Acarina, Saproglyphidae and Hymenoptera, Vespidae). Jour. Wash. Acad. Sci. (Oct.):89-92.

Krombein, K. V. 1962a. Natural history of Plummers Island, Maryland XVI. Biological notes on *Chaetodactylus krombeini* Baker, a parasitic mite of the megachilid bee, *Osmia (Osmia) lignaria* Say (Acarina: Chaetodactylidae). Proc. Biol. Soc. Wash. 75:237-250.

Krombein, K. V. 1962b. Biological notes on acarid mites associated with solitary wood-nesting wasps and bees (Acarina: Acaridae). Proc. Ent. Soc. Wash. 64(1):11-19.

Linquist, E. E. 1969. Review of holarctic tarsonemid mites (Acarina: Prostigmata) parasitizing eggs of ipine bark beetles. Mem. Ent. Soc. Canada 60:111 pp.

McDaniel, B. and E. W. Baker. 1962. A new genus of Rosensteiniidae (Acarina) from Mexico. Fieldiana-Zool. 44(16):127-131.

McDaniel, B. 1968a. The superfamily Listrophoroidea and the establishment of some new families (Listrophoroidea: Acarina). Acarologia 10(3):477-482.

Mulaik, S. 1945. New mites in the family Caeculidae. Bull. Univ. Utah 35(17):1-23.

Mulaik, S. and D. M. Allred. 1954. New species and distribution record of the genus *Caeculus* in North America. Proc. Ent. Soc. Wash. 56-27-40.

Muma, M. H. 1961. Mites associated with critus in Florida. Univ. Fla. Agr. Exp. Sta. Bull. 640:39 pp.

Newell, I. M. 1947. A systematic and ecological study of the Halacaridae of eastern North America. Bull. Bingham Ocean. Coll. 10(3):1-232.

Pritchard, A. E. and E. W. Baker. 1951. The false spider mites of California. Univ. Calif. Pub. Entomol. 9:1-94. Univ. Calif. Press, Berkeley and Los Angeles.

———. 1955. A revision of the spider mite family Tetranychidae. Pac. Moast Ent. Soc. Mem. Ser. 2:472 pp.

———. 1958. The false spider mites. Univ. Calif. Pub. Entomol. (14(3):1-274. Univ. Calif. Press, Berkeley and Los Angeles.

Robertson, P. 1959. Revision of the genus *Tyrophagus* with a discussion on its taxonomic position in the Acarina. Austral. Jour. Zool. 7(2):146-181.

Strandtmann, R. W. and G. W. Wharton. 1958. Manual of mesostigmatid mites parasitic on vertebrates. Inst. Acarology, Univ. Maryland, Contr. 4:330 pp. + vii + 69 plates.

Strandtmann, R. W. 1962. *Nycteriglyphus bifolum* n. sp., a new cavernicolous mite associated with bats (Chiroptera) (Acarina: Glycyphagidae). Acarologia 4(4): 623-631.

Summers, F. M. and E. I. Schlinger. 1955. Mites of the family Caligonellidae (Acarina). Hilgardia 23(12):539-561.

Summers, F. M. 1960a. *Eupalopsis* and eupalopsellid mites (Acarina: Stigmaeidae, Eupalopsellidae). Florida Ent. 43(3):119-138.

———. 1960b. Several Stigmaeid mites formerly included in *Mediolata* redescribed in *Zetzellia* Ouds. and *Agistemus*, new genus (Acarina). Proc. Ent. Soc. Wash. 62(4):233-247.

Summers, F. M. and D. W. Price. 1961. New and redescribed species of Ledermuelleria from North America (Acarina: Stigmaeidae). Hilgardia 31(10):369-382 + plates.

Summers, F. M. and W. M. Chaudri. 1965. New species of the genus *Cryptognathus* Kramer (Acarina: Cryptognathidae). Hilgardia 36(7):313-326 + plates.

Summers, F. M. 1966a. Key to families of the Raphignathoidea (Acarina). Acarologia 8(2):226-229.

———. 1966b. Genera of the mite family Stigmaeidae Oudemans (Acarina). Acarologia 8(2):230-250.

Treat, A. E. 1967. Mites from noctuid moths. Jour. Lepid. Soc. 21(3):169-179.

Tuttle, D. M. and E. W. Baker. 1968. Spider mites of the southwestern United States and a revision of the family Tetranychidae. Univ. Arizona Press, Tucson: 143 pp. + vii.

Woodring, J. P. 1966. North American Tyroglyphidae (Acari): I. New species of *Calvolia* and *Nanacarus*, with keys to the species. Proc. Louisiana Acad. Sci. 29: 76-84.

Woolley, T. A. 1961. A review of the phylogeny of mites. Ann Rev. Entomol. 6:263-284.

Zackhvatkin, A. A. 1941. Fauna of U.S.S.R. Arachnoidea 6(1), Tyroglyphoidea (Acari). Zool. Inst. Acad. Sci. U.S.S.R.N.S. 28 (translated by Amer. Inst. Biol. Sci.: 573 pp. + v).

Index and Glossary

A

Acarapis woodi, 203
Acaridae, 23, 24, 209, 212, 231, 238
Acariformes, 28
Acaronychidae, 286
Acaronychus, genus, 286
  A. tragardhi, 286
Acarophenax, genus, 143, 148
  A. tribolii, 142, 143, 148
Acaropsella, genus, 161
  A. kulagini, 162
Acaropsis, genus, 161
  A. sollers, 161
Acarus, genus, 241
  A. siro, 21, 238, 241
ACCESSORY SHIELDS: Paired shields outside the adanal shields in males *Boophilus* and *Rhipicephalus*.
Aceosejidae, 74
Acerodromus, genus, 78
Acer palmatum, 144
Achipteriidae, 296
Adactylidium, genus, 148
  A. beeri, 149
ADANAL PLATES: In males the paired ventral plates bordering on the median and anal plates.

ADANAL SHIELDS: Paired ventral shields near the anus in males of *Boophilus* and *Rhipicaphalus*.
Aegyptobia, genus, 166
  A. aplopappi, 166
  A. desertorum, 167
  A. nothus, 166
Agistemus, genus, 175 179
  A. fleschneri, 180
Aix sponsa, 276
Alabidocarpus, genus, 278
  A. calcaratus, 278
Alaskozetes, genus, 311
  A. coriaceus, 311
Aleuroglyphus, genus, 242
  A. ovatus, 242
Algae, 209
Alicorhagia, genus, 132
Alicorhagiidae, 132
Alliea, genus, 164
  A. laruei, 164
Allochaetophora, genus, 118
  A. californica, 118
Allochaetophoridae, 117
Allodermanyssus, genus, 60
  A. sanguineus, 60
Allogalumna, genus, 295
  A. confluens, 295
Alloptes, genus, 222
Allothrombium, genus, 8

Almond tree, 160, 198
Amaryllis, 239
Amblyomma, genus, 98, 104, 108
  A. americanum, 109
Amblyseius, genus, 73
AMBULACRAL APPARATUS: The claws and pulvillus of an ambulatory appendage.
AMBULATORY APPENDAGES: The appendages of segments III-IV used primarily for walking.
American dog tick, 105
Ameronothridae, 310
Ameroseiidae, 36
Amerozercon, genus, 82
  A. suspiciosus, 84
Amphispiza bilineata, 201, 239
Anacentrinus subnudus, 239
Analgidae, 219, 223
ANAL GROOVE: A semicircular groove posterior to the anus.
ANAL PLATE: In males the median ventral plate posterior to the median plate and surrounding the anus.
ANAL SHIELD: A sclerotized shield surrounding the anus and provided

with the setae (the para-anals and the post-anal seta).
Androlaelaps, genus, 88
  A. grandiculatus, 88
Angoumois grain moth, 142
Anoetidae, 22, 252
Anoistrocerus antilope, 227
Anoteus, genus, 254
  A. bushlandi, 255
Antelopes, 62, 107
Antennophoridae, 43
ANTERIOR PROCESS: On coxa I in *Boophilus;* the attenuated anterior extension visible in ventral and dorsal views in males.
ANTERIOR PROJECTION: The projection of the dorsal body wall at the anterior end. It may extend horizontally or may be curved ventral and be continuous with or separate from the hood.
ANTERIOR SPUR: Projection on article 1 of the palpus, directed diagonally forward; may be large (as in adult of *Ixodes auritulos*), moderate (as in the nymph of *I. sculp-*

319

CLUBBED HAIRS: Hairs which do not taper and are terminally enlarged.

CONSTRICTED: Viewed ventrally, the basis capituli may have constrictions at the sides, constant in degree in some species, absent in others.

CORNICULI: Paired horn-like structures associated with the rostrum.

CORNU (pl. cornua): Caudad projection extending from the posterior lateral dorsal angles of the basis capitulum.

CORONA: The apical portion of the hypostome which is differentiated from the remainder by having very small denticles which may be numerous or few in number.

COXA: The basal segment of the pedipalps and ambulatory appendages.

COXAE: The sclerotized portions or plates on the venter to which the legs are movably attached; their sequence from anterior to posterior is indicated by Roman numerals I, II, III, and IV.

COXAL SPURS: Large or small, long or short projections usually on the posterior margins of the coxae. Spurs may be only short, salient elevations on or near the posterior edge. The one on the median side is the internal spur, that on the outer side the external spur. When the inner spur is replaced by a lobe, it is still spoken of as a spur. In *Haemaphysalis* there may be a single spur near the middle of the posterior margin and it is denoted as a middle spur.

CRASSATE: Term used to describe the enlarged or swollen segments of leg II in some males.

CRAZED: Marked with minute cracks or crazing.

CRENULATIONS: A term applied to the transverse or diagonal rows of mile denticles found on some male hypostomes. Some variation occurs in the sizes of the individual denticles, and in whether the denticles in the crenulations are in a continuous row or separated. They are best seen under the microscope, mounted in balsam or other medium.

CRISTA: A ridge or crest.

D

DENTICLES: The recurved "teeth" on the ventral side of the hypostome. See hypostome.

DENTITION FORMULA: A "formula" indicating the number of files found on each side of the median line. Thus,

FRAME OF THE ANUS: The continuous circular or oval ring which encloses the eversible flaps of the anus.

G

GENITAL APERATURE: The external evidence of the genital organs found on the ventral median line posterior to the capitulum. Absent in nymphs and larvae.

GENITAL GROOVE: A long groove beginning near the genital aperture extending toward or to the posterior margin.

GENITAL ORIFICE: The sexual opening on the median line between the coxae.

GENITAL SETAE: Paired setae on or in close proximity to a sclerotized genital shield.

GENITAL SHIELD: A sclerotized shield covering the genital orifice in the females. It may be provided with an anterior hyaline extension which functions as a chute during oviposition.

GENITAL-VENTRAL SHIELD: The compound shield, bearing more than one pair of setae formed by the fusion of the genital shield with part of the ventral shield.

GENU: A segment of the pedipalp and legs between the femur and tibia.

GNATHOSOMA: The region of the body that encompasses the area beyond the first pair of legs including the chelicerae, known as the head region or the region containing the mouthparts.

GOBLETS: Within the spiracular plate, small structures resembling pores. These pores may be large and few or small and numerous.

GRANULATIONS: Irregular elevations on the surface of the integument in adults of *Otobius,* in contra-distinction to tubercles in *Antricola* and mammillae in *Ornithodoros,* and "elevations" or wrinkles in *Argas.* Also used in describing the very small elevations, similar to micro-mammillae, found in some species.

GROOVES: Lineal depressions or furrows, mainly on the ventral surface. Their depths and widths much influenced by the degree of engorement.

H

HAIRS: Hairs on the basis capituli, scutum and legs are usually simple and tapering but in some species the hairs on the scutum may not taper but may end abruptly; on the tarsi some hairs may be faintly plumose or forked.

tribution of the punctations are often useful.

VENTRAL SCUTES: Chitinous thickenings of the ventral surface of the festoons which may be very distinct and protruding, faint or absent. They may also bear nubs.

VENTRAL SHIELDS: On the ventral surface in males of *Rhipicephalus*, *Boophilus* and *Hyalomma*; the irregularly shaped shields usually attached in front and free behind; designated as adanal and accessory plates.

VENTRAL TARSAL SPURS: Spurs on the ventral side of the tarsus designated as terminal and subterminal ventral spurs.

Vespacarus, genus, 223, 224
V. histrio, 224
Vespid wasp, 251
Vidia, genus, 224, 225
V. cooremani, 225
Vigna sesquapedata, 136

NOTES

NOTES

NOTES

NOTES